Gas Lift Systems
Design & Optimization

GAS LIFT SYSTEMS DESIGN & OPTIMIZATION

A MODERN MODELING APPROACH

MANICKAVASAKAN S. NADAR

PB
PENNWELL BOOKS

Disclaimer

The recommendations, advice, descriptions, and the methods in this book are presented solely for educational purposes. The author and publisher assume no liability whatsoever for any loss or damage that results from the use of any of the material in this book. Use of the material in this book is solely at the risk of the user.

PennWell Books, LLC
10050 E 52nd Street
Tulsa, Oklahoma 74146 USA

866.777.1814
sales@pennwellbooks.com
www.pennwellbooks.com

Publisher: Matthew Dresher

Library of Congress Cataloging-in-Publication Data Available on Request.

Printed in the United States of America

1 2 3 4 5 26 25 24 23 22

Contents

Figures

Tables

About the Author

Manickavasakan S. Nadar (popularly known as *Manickam*) is a well-known expert, specializing in the field of petroleum production engineering and operations, artificial lift, system modeling, and optimization. With over 30 years of experience, he is a senior consultant and professional trainer in petroleum engineering.

Manickam has contributed to the improvement of efficiency in the design, operation, and troubleshooting of large-volume artificial lift operations, resulting in significant improvement of production and revenue. He had an illustrious career with Dubai Petroleum Company (then a business unit of Conoco Inc.), handling lead roles in production operations, well completions, artificial lift, asset modeling, and optimization. Later, he worked as a Principal Petroleum Engineer with Edinburgh Petroleum Services Ltd. (a Weatherford company), in Edinburgh, UK, and was responsible for development and implementation of optimization projects of many giant oilfields in the regions of Asia and the Middle East. He had also worked as a Senior Production Technologist with Margham Dubai, UAE. Before moving to the oil and gas sector, he gained a wealth of knowledge and experience in process operations at Asia's largest single-stream Ammonia plant, SPIC Ltd., in Tuticorin, India.

A strong believer in human creativity and use of technology, Manickam has helped clients move their offline production optimization to real time, sustaining huge benefits of optimization, including increased profits, reduced downtime, and energy savings. Manickam's contribution to gas lift engineering and operations using a lesser number of valves selected based on dynamic testing has improved the reliability of gas lift designs and led to cost reduction. His contribution to the improvement of network optimization software solutions, especially for well gas lift design and gas compressor modeling is noteworthy. His work has led the operation of many old fields with high water cut to become very economical.

During his career with Conoco and Weatherford, Manickam was actively engaged in mentoring his colleagues through in-house training sessions. As an independent consultant, he is now associated with development and delivery of training through PetroSkills®, serving the oil and gas industry across the globe. He has developed training content in his areas of expertise for many training courses. His focus is on achieving excellence through system optimization, a collaborative work environment, avoidance of redundant work, and the proactive contribution of individuals.

Manickam has a number of publications to his credit. He has made presentations in many Society of Petroleum Engineers (SPE) meetings and other international and regional conferences, on various topics related to artificial lift, well surveillance, and asset optimization. He served as guest trainer in artificial lift economics in SPE conferences in Cairo and Ahmedabad. He has won several awards from Conoco for his creative and innovative work and teamwork skills.

Manickam holds a BS degree in Chemistry from Madurai University, India and a degree in Chemical Engineering from Institution of Engineers (India). Apart from petroleum engineering, his passion includes Indian classical music and charity work.

Author's Note

HAVING WORKED WITH SEVERAL LARGE GAS LIFT SYSTEMS FOR THE PAST 30 YEARS, I am certain beyond doubt that there is always scope for improvement in gas lift operations. The quest for excellence in applying engineering investigations and setting up best practices with this versatile artificial lift system is ripe with rewards. This keeps engineers engrossed and constantly seeking opportunities for progress.

In the past three decades, as computers took up desk spaces, we found novel ways of refining gas lift technology by performing more field tests and analyses than ever before. This helped transform the *art* of gas lift to more of a *science.* Then came the communication explosion that facilitated sharing of these wonderful findings across the globe. As a result, the industry started applying engineering recommendations based on actual tests, investigations, and analysis, as opposed to the rules of thumb that had been previously followed. These developments allowed gas lift designs to become more accurate.

The spirit of growth did not stop there. The young and aspiring engineers from operating companies wanted to optimize their asset performance in a holistic manner, as opposed to merely solving individual well problems. This lit a new path in the industry. The service industry and operators collaborated in a big way and focused on the optimization of assets and networks, paying close attention to the performance characteristics of *every component of the network,* not just wells. Thus, the journey of network modeling and optimization took off. Time and again, case studies have proved the positive economic benefits of the gas lift system optimization approach. A more creative approach to system optimization not only opens up new possibilities, but also turns existing constraints and inefficiencies into profit-making opportunities.

What's more, moving to real-time surveillance and optimization yielded far more benefits than offline optimization, as this new methodology helped minimize downtime and production losses. The real-time optimization (RTO) application is able to readily identify the component or area of the field that is underperforming—making it possible to start appropriate remedial action right away. This is one of the key benefits of the "digital oilfield."

Always open to exciting innovation, the energy industry now uses the new data science—big data analytics, machine learning, and artificial intelligence—that is ushering in significant changes in the way the assets are monitored and optimized. Believe it or not, these powerful analytical tools possess human-like cognitive skills and are capable of modifying optimization algorithms by self-learning and training. As a result, more workflows get automated, downtimes are reduced, and efficiencies are further improved upon.

After successful implementation of the field operation center (FOC) for every asset, the industry has moved on to digital data integration for achieving inter-asset optimization through a real-time operations center (RTOC). The RTOC (Production) is extremely powerful in identifying synergies and maintaining high levels of efficiencies among many assets that are connected to each other with integrated production and export targets.

At every step of this innovative journey, the results kept improving the quality of oil and gas operations. Production was increased, and costs were reduced. Collaboration paid off as exciting solutions emerged.

Having had the privilege of being a part of this revolutionary industry in various roles, I have written this book as an attempt to capture how we optimize gas lifted fields using the modeling tools and communications technology available to us now. This book is intended for engineering and operations staff of operating and service companies, who work together to use technology intelligently to improve asset profits. All the information discussed is fully field-based and result-oriented. This book aims to provide you with the practical wisdom to embark on field gas lift system optimization using modern, pragmatic, applied, and proven methods.

Foreword
Paul M. Barry
Course Director, Production and Completions Engineering Division, PetroSkills LLC

SUCCESS IN OPTIMIZING A GAS LIFT SYSTEM to achieve both production and cost management goals flows from the sound application of informed technical reasoning and assumptions, supported by local knowledge based on technical, logistical, and commercial realities. That sound application originates in superior engineering design and then must be followed by the operations execution of a well-developed gas lift well completion and system optimization program.

Throughout his impressive career, **Manickam Nadar** has excelled in necessary strategic thinking and the resultant skilled application of both proven and emerging artificial lift principles and practices. Being an artificial lift expert, he is a leading industry consultant in gas lift and other artificial lift technologies. In the 15 years I have known Mr. Nadar, his work as oil and gas engineer, consultant, and subject matter expert for major oil companies, government national oil companies, independents, training organizations, and oilfield service companies has been fully professional, knowledgeable, precise, detailed, and complete. Underlying all these attributes is a full commitment to safe operations and the means and management vision to achieve such.

Gas Lift Systems Design & Optimization: A Modern Modeling Approach is a specialized professional book that covers all aspects of production optimization of gas lift systems. The book captures the depth and breadth of Mr. Nadar's comprehensive experience, bringing both leading-edge insight and practical methods to industry engineers and managers seeking the latest in optimizing gas lifted assets. Inherent in his approach in this text is the logic of a holistic design concept presented by Mr. Nadar that assures a comprehensive system methodology. The importance of blending a thorough understanding of fluid properties, fluid inflow capacities (oil, gas, water, condensate), gas lift design options, gas lift supply and production gathering facilities, operation of processing equipment, well and field management strategies, and several related parameters is demonstrated in this solid approach, which enables engineers to apply and achieve excellence in various key performance indicators.

My experience working side by side with Mr. Nadar in many oilfield production operations scenarios assures confidence in the precept that application of the practical principles and methods contained within the text will lead to proper choices made by engineers to maximize value from the application of gas lift technology in their producing assets. Newer, emerging technologies such as artificial intelligence, data mining, and digital twin, along with making use of the exponential improvement in modeling and optimization, combine to provide the reader with a Best-in-Class cookbook approach to designing and operating gas lift systems.

It is often proposed that gas lift is the most forgiving of all the family of artificial lift technologies. This principle is regularly observed as being accurate, and thus, gas lift well performance surveillance and precise decision making can provide a great understanding of what adjustments might be made to improve gas lift system performance. Methods explained in detail in the book provide a professionally written checklist of gas lift interrogation bullet points to adopt.

A point of emphasis often presented as a consultant is that oilfield operations management at various levels and disciplines must provide the means for the execution of well-developed ideas. Simply put, brilliant ideas disappear if "action," guided by commercial value assessment, does not follow a decision to consider a change when needed in the engineering, process, or equipment. This book provides the underlying basics for achieving quality and maximizing profitability in the performance of both individual gas lift completions as well as the overall field network. It also provides confidence that the practical implementation of principles of gas lift system optimization presented by Mr. Nadar will lead to positive results in any asset.

At some point, many oilfield engineers compile and assemble their career experiences to collect and organize their lifelong compendium of knowledge, know-how, and insights. Many do so just for the peace of mind achieved in "putting it all together." Others do so simply as part of their job responsibility to gather local knowledge in a particular company or geographical oil and gas basin. Regarding Mr. Nadar's approach to system modeling and optimization, the material contained in the book provides a solid, leading-edge, contemporary, technically accurate, and precise summary for engineers at any career stage—thereby enabling them to improve upon their understanding and application of artificial lift to maximize production and achieve commercial success for their oil and gas industry organization.

Paul M. Barry
September 28, 2021

Acknowledgements

This book is the culmination of several years of my work in the oil and gas industry.

I am indebted to my senior colleagues who helped me explore the creative and innovative dimensions of engineering during my early days at Conoco. True engineering investigation is no easy feat—it encompasses a lot of reading, data analysis, focused discussion, and dedication. My peers and seniors have tremendously aided in establishing my foundation, and for that, I shall always be grateful.

In any optimization project, consultants come across a few highly dedicated persons from the operating company, who provide the spark for tough projects and refuse to settle for anything less than the right solution. These are the real heroes who truly help and guide consultants' work to meet, or even exceed, their expectations! I have had the privilege of working with some of them and tackling their challenges during my career. I owe a lot to these individuals: Arun S. Kallal from Dubai Petroleum, Dr. Fathi Shnaib from Petrofac, Pavan Aggarval from ONGC, Chandran Peringod from Shell, and Ahmed Bin Amro from ADCO, to name a few.

Starting with my SPIC and Conoco days, mentoring has been a delightful experience for me, and I am proud to see that many people I coached have become experts in various areas of petroleum and process operations. There is no greater joy than this!

The team at PetroSkills has been another exciting source of inspiration. My contribution to the curriculum development for some of the petroleum engineering short courses and the training instruction to the industry colleagues has been a rewarding experience. I express my sincere thanks to my colleagues at PetroSkills, especially Paul Barry, one of the course directors of the Production and Completion engineering discipline, who insisted I write this book and also gave it its title.

I thank Kumud Ranjan and Vibhuti Dwivedi, petroleum engineering consultants for their enthusiastic assistance in reviewing the material of chapters 8 (Data Analytics) and 9 (Data Integration).

I enjoyed my interaction with the officials at PennWell Books, especially Matthew Dresher, for his valuable assistance in defining the scope of the book and formatting its contents. He has been very helpful and positive from the very beginning. I am very grateful to the senior professional who reviewed the draft version and made suggestions that have helped improve the focus of the book significantly.

Looking back, I thank my mother and father who set some personality traits of a very high order for me to achieve in life, some of which I am still struggling to accomplish. But that has made my journey full of purpose and happiness. I am very thankful to my wife and children who made many sacrifices that made it possible for me to spend a lot of time building my career. Without their perseverance and support, I could not have made it here.

Finally, I thank God for everything that has come my way in life.

With sincere thanks,
Manickam

Outline of the Book

This book presents efficient and practical ways of designing and managing gas lifted network systems from a field perspective. Maintaining peak efficiency in the operation of gas lifted assets is all the more important in today's environment. Gas lift system optimization methods described herein will establish that superior economic benefits can be achieved if each component of the production network is reviewed and optimized, instead of optimizing the well performance alone. The book will demonstrate how this can be achieved at a practical level. The book will also highlight the use of artificial intelligence, machine learning, and data science technologies that are available today to take asset optimization to greater heights and extend the benefits across various assets.

The book covers the following:

- How to improve performance of the production network by modeling every component of the network, along with the wells, and using the model-based optimization to run gas lift operations at peak efficiency, during normal operations as well as process upsets.
- How to make use of powerful real-time data visualization to identify underperforming components of gas lift networks and a real-time optimization system to maximize revenues and minimize operating and maintenance costs.
- How to utilize data transparency and transform asset operations to a collaborative working environment that enhances teamwork efficiency and helps an organization to achieve its key performance indicators (KPIs).
- How to optimize gas lifted wells, with special emphasis on unconventional reservoirs and deepwater gas lift operations.
- How to apply big data analytics, neural networks, and machine learning technologies to improve equipment reliability and plant availability. The goal is to spot areas of sub-optimal performance and correct them before incurring loss of production.
- How to integrate various assets inked to the hydrocarbon chain to identify synergies and promote inter-asset optimization. The real-time operations center (production)—RTOC (P)—can help integrate oil and gas operations within an organization or across multiple organizations in a country.

Chapters

1. Introduction to Gas Lift Network Optimization

This chapter introduces the use of well performance for gas lift optimization of a single well. Having discussed multi-well gas allocation and gas lift strategies, the concept of network optimization is introduced. The importance of the various components of the network such as pipelines, gas lift compressors, and separators is highlighted. This topic is an area that operators generally overlook, and, as a result, this omission causes serious system inefficiency, which is often expensive to correct.

2. Network Optimization Becoming a Reality

The foundation for establishing a versatile network system optimization is explained in this chapter. The factors to be considered during the evaluation and selection of the system optimization software are described in detail. The various steps involved in the implementation of system optimization have been explained systematically.

3. Modeling of Network Components

Details of the modeling of various components of gas lift production networks are presented. This includes fluid modeling and modeling of wells, pipelines, rotating equipment, and various other components of the surface network. Application of these concepts would help the reader fully understand the concepts of engineering modeling and gain the confidence to implement production optimization at the system level for any asset, of any size and complexity, irrespective of the software tools chosen.

4. Running the System Optimization Model

Finer details for running the model such as setting objectives and constraints for history match and optimization case studies are covered in detail. Many optimization scenarios of gas lift system optimization have been illustrated. This will help the user get the maximum value out of the system optimization model for solving day-to-day problems and evaluating the merits of project proposals.

5. System Modeling and Optimization Case Histories

Details of three real-life success stories are presented. These are case histories of design and implementation of gas lift system optimization in large assets. Also included are the practical challenges that were encountered during the implementation and the solutions provided in the projects. Factors that are critical for the successful implementation of gas lift system optimization are highlighted.

These should give the practical wisdom that is needed for the engineering and operation teams of the operator, as well as the service consultants.

6. Real-time Optimization

This chapter emphasizes the value of bringing data to desktop and implementing real-time optimization in assets. The need for and benefits of real-time surveillance and optimization of large production networks are presented. The intelligent systems that are used for implementation of real-time optimization (RTO) systems are described. How various KPIs are set, monitored, updated, and displayed in the real-time environment are captured. Sample displays showing various ways of visualization of results have been included.

7. Optimization of Unconventional and Subsea Gas Lift Wells

This chapter of the book describes the production optimization and artificial lift strategies of unconventional wells and deep water gas lift wells. New techniques of gas lift design & optimization of unconventional wells are covered in detail with supporting case histories. The challenges associated with deepwater single-point and riser base gas lift are highlighted. Gas lift optimization solutions have been presented for these cases.

8. Data Analytics and System Modeling

The eighth chapter takes the reader to the world of big data analytics and explores the impact of the digital transformation that is now happening in the oil and gas industry. The author explains how the advanced tools of machine learning, neural networks, and artificial intelligence can bring a sea change in the way the oil and gas industry will work in the future. The chapter explains how plant availability and uptime for critical machineries can be improved through predictive and prescriptive maintenance strategies using data analytics and machine learning models. How problems in gas lifted wells and networks can be identified is explained with sample data.

9. Digital Integration of Assets

The concluding chapter highlights the benefits of linking many producing assets and "customer units" through the concept of the RTOC (P). While field operations centers align the engineering and operations within each asset, the RTOC helps manage the inter-asset collaboration to ensure improved efficiency levels for all of the units. The digital integration of assets uses the physics-based and analytical models and identifies synergies across various operating units. The concept can be extended beyond a single organization to integrate upstream, midstream, and downstream operations of the hydrocarbon chain.

1

Introduction to Gas Lift Network Optimization

Production optimization of a gas lifted field always begins with wells. It is necessary to understand the heart of a well optimization—the gas lift response curve—generated for every well, using a well model. The "optimum" and "maximum" lift gas for a well and its current operating slope can be understood from the analysis of the gas lift response curve. Gas lift distribution to the wells in a field with a wide range of production capacities is always a challenge. The changes that happen to the wells and oilfield equipment over time represent another issue that would impact the gas distribution process. The ultimate solution for the optimization problem is to shift the paradigm from well gas lift optimization to "network production optimization."

This chapter discusses the following topics:

- Gas lift optimization using a well model
- The gas lift response curve of a well
- The concept of incremental slope
- Multi-well gas allocation
- Use of gas lift injection hit-lists
- Full network optimization

1.1 Well Gas Lift Optimization

1.1.1 Types of Production Optimization

In the oilfield context, optimization would mean either *increase production* (i.e., increase the flow rate from wells, maximize the reservoir production, or lower the mean failure rates) or *lower the costs* (i.e., reduce downtimes, maximize human resource potential, and do more with less).

There is a tendency to classify the areas of optimization based on where the tools are deployed or the parameters gathered from:

- **Well optimization** provides intelligence at the wellsite and desktop by installing specific controllers for each form of lift and monitoring with desktop software with an advanced analytical tool.
- **Field optimization** analyzes a total field's effectiveness using software tools that optimize well service and improve operational efficiency; this allows better priority of tasks and maintenance of well records.
- **Reservoir optimization** allows real-time remote monitoring of pressure, temperature, and seismic and other parameters. This can alter the reaction of each well's production to the changes in the reservoir.

The overall objective of optimizing wells and field is to improve the ultimate recovery from the reservoir in an efficient manner. In other words, the production and lift technology employed should assist and become a part of the overall reservoir optimization of the asset.

1.1.2 Optimization and Asset Management

Of all forms of optimization, the one that will make a big difference in terms of ultimate hydrocarbon recovery from the reservoir is asset management, which will cover the entire life cycle of the asset. But throughout the life of the field, there are also many short-term optimization activities that help improve efficiency of operations and increase net revenue from the asset. These are performed with the help of conventional and emerging technologies.

The asset optimization opportunities fall under four broad categories:

1) *Control and automation*—Instantaneous loop, ranging from minutes to hours
2) *Production optimization*—Short term loop, covering from days to weeks
3) *Reservoir management*—Medium term loop, extending from months to years
4) *Asset management*—Long term loop, covering the life cycle of the asset, i.e., for many years

The first two can be categorized under *fast loop optimization,* whereas the last two fall under *slow loop optimization.* Well and network optimization, performed either online or offline, falls under the fast loop category.

Fig. 1-1 shows how these four optimization loops contribute to improved operations of the asset.

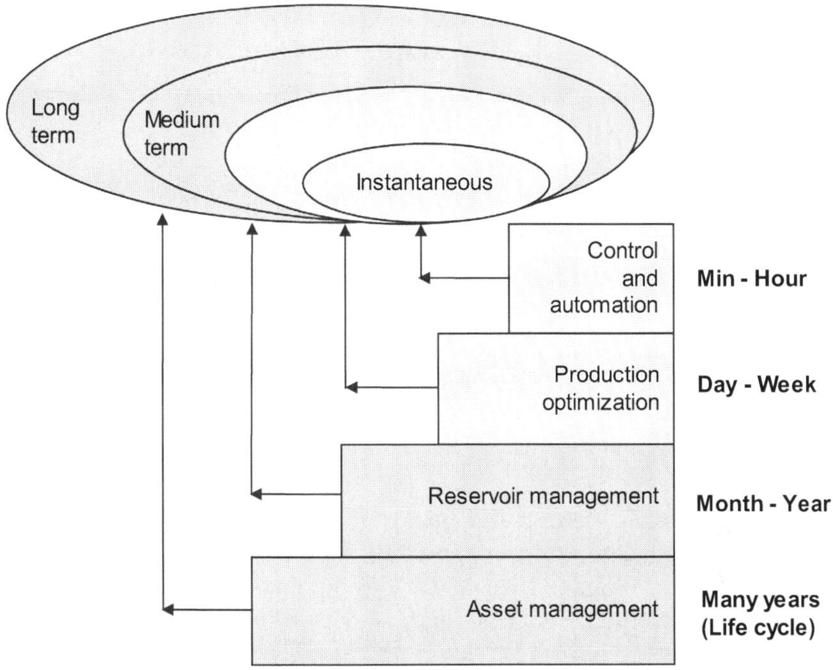

Fig. 1–1. Fast and slow loop optimization

1.1.3 Types of Gas Lift Optimization

With specific reference to gas lifted wells, three levels of gas lift optimization can be identified:

- **First level—Individual well performance optimization:**

  This is aimed at maximizing productivity of individual wells. The areas of focus would be well completion, stimulation, artificial lift designs, and trouble-shooting. This does not focus on resource distribution.

- **Second Level—Allocation of resources to maximize production from a 'group' of wells:**

  This can take into consideration of costs as well as revenue in deciding where to allocate resources. For example, this would include a plan for lift gas allocation and electrical submersible pump (ESP) power distribution for a group of wells.

- **Third Level—Optimizing overall performance of a production system**:

  This is the true optimization problem, taking into account various possibilities for revenue generation, considering different constraints present in a production network. This is usually performed offline and implemented periodically. Once the offline execution is successful, the next step is achieving real-time optimization.

Any organization would start gas lift optimization with the first level—i.e., optimizing the wells. From first level, operations would move to second, and finally to third level.

1.1.4 Modeling of a Gas Lifted Well

The first step of gas lift optimization is to create a well model. The model may be built before drilling the well, but should be updated with the actual well completion and production information after the well comes on stream and stabilizes. A model calibrated with the stabilized production test data can be used for further investigations. These investigations help operators in optimizing wells in the field.

For about four decades, nodal analysis has been extensively used to design, analyze, and optimize wells. This includes naturally flowing and artificially lifted wells.

In simple terms, three different sets of data are required to create a well model:

- **Inflow**

  Inflow defines the flow potential of the well to deliver fluids from the reservoir to the wellbore or the energy required to deliver the fluids from reservoir to wellbore. Details such as reservoir pressure and temperature, rock properties, fluid behavior, and near-wellbore restrictions are considered to arrive at the inflow description of the model. The inflow is mainly controlled by reservoir properties and type of well completion at sand face. If the reservoir properties are not readily available, the engineer would estimate the properties using offset well data. If the well is already on production, using the well test data, an empirical approach can be used to define inflow. Based on the inflow information, the production rate possible from a well for a specific drawdown would be known.

- **Outflow**

  Outflow defines energy required to deliver the fluids from wellbore to surface facilities. Details of the vertical completion (tubing / casing sizes, restrictions in tubing, wellbore deviation, and artificial lift equipment) and surface piping are included in the outflow description. The pressure drop in the outflow system is more readily calculated than inflow. The model will use a multi-phase flow correlation suitably selected for the well conditions to calculate the pressure drop in the outflow section. The correlation selection depends on many factors such as production rate, size of production tubing, gas-liquid ratio, water cut, presence of emulsion, and well deviation.

  The suitability of the selected correlation can be verified using flowing gradient surveys from the same well or a similar well flowing from the same formation.

- **Pressure-Volume-Temperature (PVT)**

  Fluid properties (PVT properties of various fluids handled by the wells) are incorporated in the well model. As the fluids flow from reservoir to surface separator, the pressure and temperature will change. The behavior of fluids under these changing conditions should be captured accurately. For fluid modeling, a representative piece of PVT data for the current reservoir conditions is preferred. If this is not available, the PVT analysis reports of the reservoir fluids from old wells may be used. Offset well data and knowledge of the reservoir operations are needed to get the fluid modeling done as accurately as possible.

The result of the nodal analysis is a plot of inflow-outflow curves, intersecting at a point called the "operating point." The operating point represents the current flowing condition of the well. It shows the current flowing bottom-hole pressure (FBHP) and the liquid production rate (**Fig. 1-2**). The "bottom-hole" refers to the mid-perforation depth or mid-horizontal section of the well.

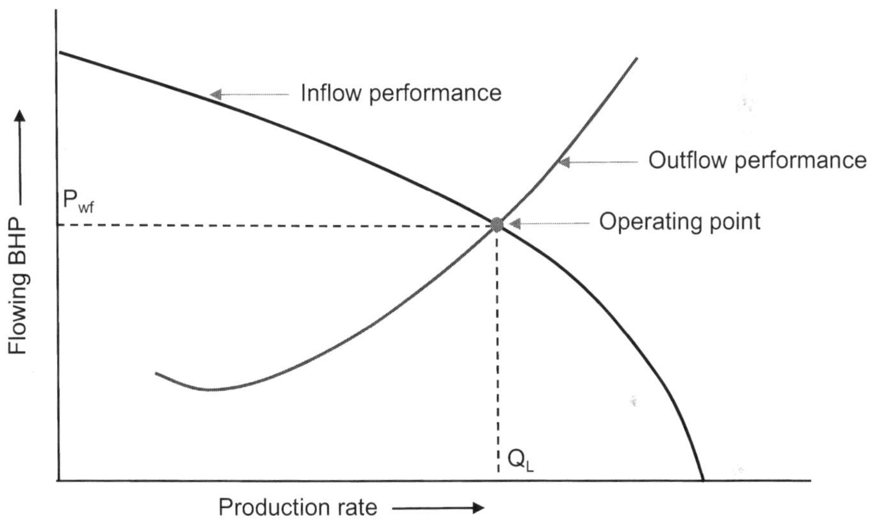

Fig. 1–2. Well operating point

A calibrated model can be used to investigate several scenarios, i.e., to understand how the well will perform if inflow or outflow parameters change.

1.1.5 Gas Lift Response Curve

While designing gas lift for a well, the designer performs several nodal analysis runs to confirm that the gas lift design finalized for the well meets the design objectives. The primary objectives of a good gas lift design are listed below:

- Gas lift the well from the deepest possible depth (maximize hydraulic efficiency)
- Maintain single-point injection with gas lift (avoid multi-point injection)

- Maximize production gain from the well within economic limits of gas lift injection rates
- Optimize tubing size for the production rate expected: very large tubing leads to flow instability problems; very small tubing leads to excessive frictional pressure loss in tubing

The designer will determine mandrel spacing, unloading valve port size and pressure settings, location of operating point, sizing of the operating valve, and the lift gas injection rate to meet these objectives. After the initial design, however, it is possible that the parameters such as gas lift operating depth and lift gas rates may be changed, based on actual flowing conditions of the well and gas lift source available at the well location. The mandrel spacing remains unchanged.

The common gas lift configuration uses gas injection down the casing-tubing annulus and production up the tubing. High-pressure gas injected into the production casing annulus flows through the operating gas lift valve into the production tubing. Knowing the surface pressure and properties of the lift gas and the sizes of the casing and tubing, the nodal analysis model will determine the pressure at the injection depth in the annulus. This is to ensure that sufficient pressure is available to inject lift gas into the tubing at the designed depth. The pressure differential required for injecting the required volume of lift gas rate though the operating valve is also modeled.

The nodal program models the impact of gas lift, i.e., increase in gas-liquid ratio, reduction in density, gas expansion, and increase in velocity. Hydraulic calculations are performed by the multi-phase pressure-drop correlation selected by the user. Starting with the prescribed flowing wellhead pressure, a flowing pressure traverse is established for the given gas lift injection rate, arriving at a flowing bottom-hole pressure for these conditions. The drawdown so established determines the production rate from the well, as shown in **Fig. 1-3**.

In an existing gas lifted well, the two parameters that can be altered to control fluid production are the gas lift injection depth and gas lift injection rate.

- **Gas Lift Injection Depth**

  The gas lift injection depth is prescribed by the gas lift design of the well. The operating valve is positioned at this depth. Once the well comes online, the gas lift depth can be verified with downhole pressure and temperature surveys. A gas lift redesign will be required to change the injection depth; this may involve changing gas lift valves in the well completion through wireline intervention.

- **Gas Lift Injection Rate**

  After defining the gas lift depth, the gas lift rate is set at a recommended value, based on the production potential of the well and lift gas availability in the field. The gas lift injection rate can usually be varied within limits.

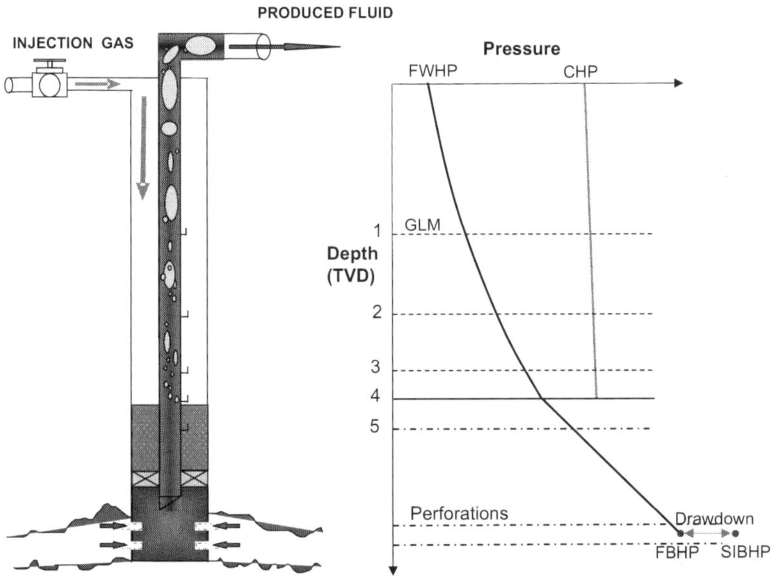

Fig. 1–3. Gas lift flowing gradient in a well

The operating point analysis can be repeated for different gas lift injection rates, *assuming* a constant flowing wellhead pressure and unlimited supply of lift gas. The inflow parameters are kept constant for this analysis. The operating points obtained for different gas lift rates are marked in **Fig. 1-4**.

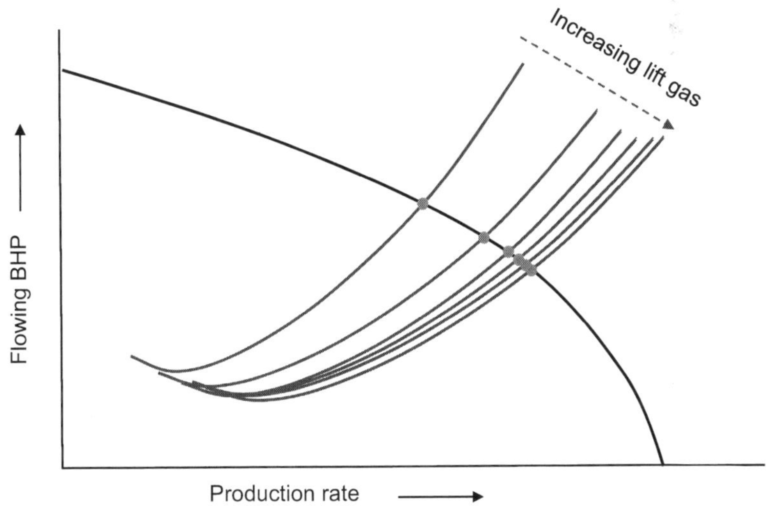

Fig. 1–4. Operating points for different gas lift rates

The results of operating points obtained for various gas lift rates can be used to construct a plot of oil production rate versus gas lift injection rate. This plot is called a *gas lift response curve* or a *well performance curve.*

A typical well gas lift response curve so obtained is shown in **Fig. 1-5** :

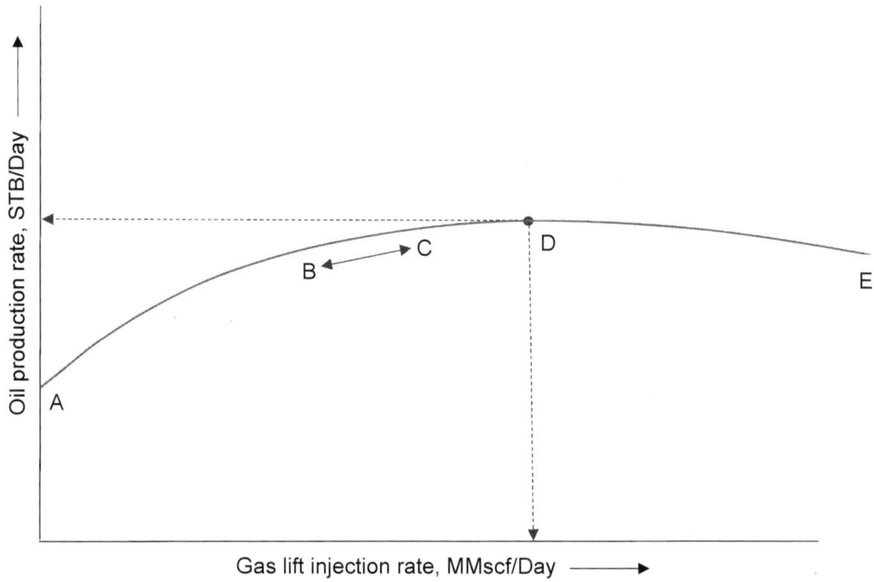

Fig. 1–5. Gas lift response curve

The shape of the curve for every well is different and is controlled by several factors such as inflow parameters, tubing size, water cut, formation gas-oil ratio (FGOR), flowing wellhead pressure, and the multi-phase flow correlation used for the hydraulic calculations.

The curve ABCDE in **Fig. 1-5** represents the performance curve of a well that can flow without gas lift. Point A represents the production with no gas lift injection. Oil production increases with lift gas injection, but beyond a point, the incremental oil gain declines with further increase of lift gas. Point D represents the maximum production possible from the well under current operating conditions. The response curve stays flat for a while and then starts decreasing. Section DE represents the area of over-injection.

As high pressure gas is not free, it will not be economical to operate any well at point D. The well should be operated at an *optimum rate* (section BC). The optimum rate depends on the cost of lift gas and back pressure effects on the production system.

For some wells, the gas lift response curve generated by the multi-phase flow correlation does not match with the well response. In such circumstances, stabilized production rates from the well, measured at different gas lift injection rates, will be used to construct a gas lift response curve or a well performance table for the well.

A gas lift response curve is generated for every well in the field using the well flow model. This is the starting point of gas lift well optimization. The next important step is determining the optimum gas lift rate.

It should be noted that the two assumptions of constant flowing wellhead pressure (FWHP) and unlimited supply of lift gas are incorrect. Thus, the true optimization should take these factors into consideration.

1.1.6 Incremental Slope

The term *incremental slope* usually refers to the incremental oil production rate gained from the well for the last 1 MMscf/D of lift gas injected. If a well produces 1000 BOPD for a gas lift injection rate of 3.0 MMscf/D, and if the production rate increases to 1040 BOPD for a gas lift rate of 3.5 MMscf/Day, the incremental oil slope = dQ_o/dQ_{gi} = 40/0.5 = 80 BOPD/last 1 MMscf/D injected.

The gas lift response curve for a gas lifted horizontal well from a field, generated using a calibrated well model, is shown in **Fig. 1-6**. The most recent well test rate is also marked.

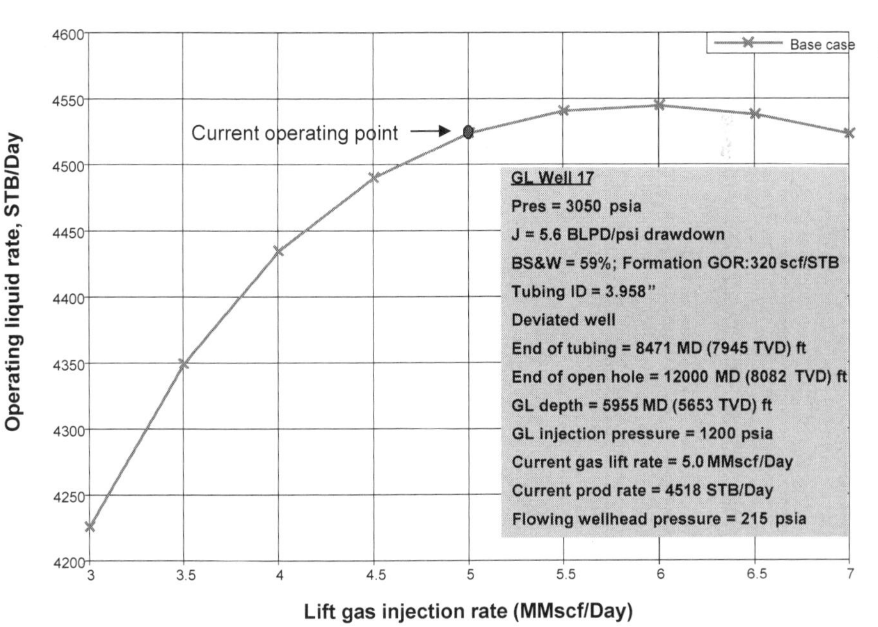

Fig. 1–6. Gas lift performance curve from a well model

The production rates read from the graph are tabulated in **Table 1-1**.

Table 1–1. Well gas lift performance curve data

Gas-lift injection rate, MMscf/Day	Liquid production rate, STB/Day	Oil production rate, STBO/Day	Incremental slope, STBO/MMscf	Lift gas cost, $/Day	Oil revenue, $/Day	Net profit, $/Day
3.0	4225.0	1732.3		1500.0	51967.5	50467.5
3.5	4350.0	1783.5	102.5	1750.0	53505.0	51755.0
4.0	4436.0	1818.8	70.5	2000.0	54562.8	52562.8
4.5	4492.0	1841.7	45.9	2250.0	55251.6	53001.6
5.0	4525.0	1855.3	27.1	2500.0	55657.5	53157.5
5.5	4540.0	1861.4	12.3	2750.0	55842.0	53092.0
6.0	4542.0	1862.2	1.6	3000.0	55866.6	52866.6
6.5	4534.0	1858.9	–6.6	3250.0	55768.2	52518.2
7.0	4522.0	1854.0	–9.8	3500.0	55620.6	52120.6

At the current lift gas rate of 5.0 MMscf/D, the incremental slope of this well is 27 STB Oil/Day for the last 1 MMscf/D injected; it becomes almost zero at a gas lift rate of 6.0 MMscf/D. This means the production rate has reached its maximum value. At 6.5 MMscf/D the slope is negative. Addition of lift gas leads to loss of production.

The gas lift response curve is plotted along with the incremental slope in **Fig. 1-7**. The plot confirms the incremental oil slope becoming zero at a gas lift rate of 6.0 MMscf/Day. The optimum gas lift rate should be less than this value.

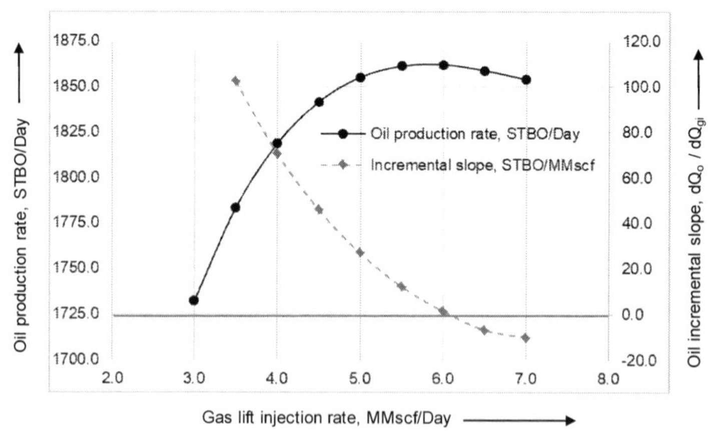

Fig. 1–7. Gas lift response—incremental slope

However, this approach treats as if lift gas is free, which is not true.

The most efficient injection rate is the point at which the incremental revenue is equal to the incremental cost of injection and not simply when production is maximized.[1] Thus, the most efficient total gas-liquid ratio (TGLR) can be established only after considering the price of oil and the cost of lift gas.

The most economical TGLR is achieved when the profit obtained from lifting a quantity of oil by a unit increment of gas is just equal to the cost of that unit of gas.[2] In other words, if the last 0.5 MMscf/D of lift gas that costs $250 creates a net oil revenue of $250, then the maximum gas lift rate has been reached.

Assuming a price of $30/BBL for oil and a cost of $500/MMscf for lift gas, the *net profit* for various gas lift rates can be calculated in $/Day, as shown in **Table 1-1**.

With these parameters set for oil price and gas cost, the net profit from this well reaches maximum at a gas lift rate of 5.0 MMscf/Day, and not at 6.0 MMscf/Day. This can be observed by plotting a gas lift response curve along with the net profit (**Fig. 1-8**).

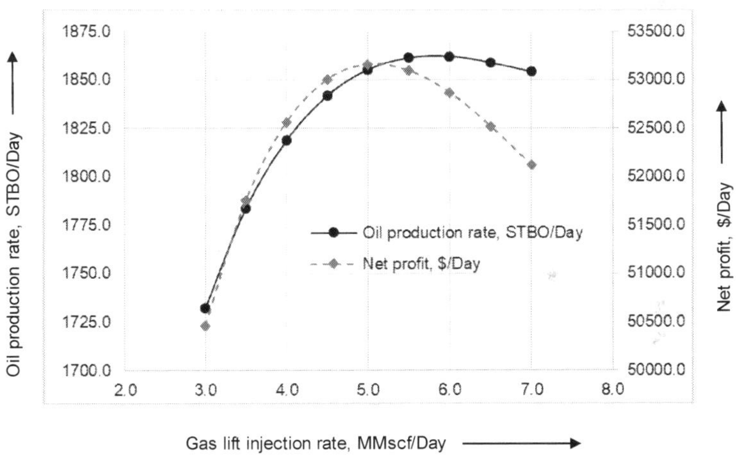

Fig. 1–8. Gas lift response—net profit

While determining the optimum gas lift rate for a well in a network, apart from oil price and cost of lift gas, the well performance and interaction with other wells connected to the manifold should be taken into account.

Sometimes a decision has to be made as to whether the field should make use of excess gas available from an external source, at a cost. Suppose lift gas is available for import from an external source at a cost of $5,000/Day for 5 MMscf/D of lift gas. The operator should decide if this import will be profitable or not.

- If an increase of 500 BOPD is obtained by using the extra lift gas of 5 MMscf/Day, with an oil price of $30/BBL, the additional revenue generated is $15,000/Day. In this case, it is worthwhile to purchase the extra lift gas and use it.

- If the increase in production is 150 BOPD only, the additional revenue from oil production ($4500/Day) would not compensate for the cost of lift gas purchased. This case does not favor using the additional lift gas in the field.

Sometimes, depending on the volume of gas being circulated already in the network, additional lift gas may not increase the profit even if it is available free of cost.

The concept of incremental slope is useful for determining the expected increase in production by adding lift gas. It is also used for gas allocation in a field. This concept makes operations focus on the benefit from an increase of oil production as a return on the additional investment needed to increase the volume of lift gas.

1.2 Multi-well Gas Lift Allocation

This comes under Level 2 gas lift optimization. The maximum available lift gas rate is usually known from the installed compression capacity. The production optimization team also knows the incremental slope the wells are operating at, based on their Level 1 optimization experience. Armed with this information, the team allocates the available gas to the wells to maximize oil production.

1.2.1 Lift Gas Allocation during Normal Operations

Wells have different productive indices (PI), water cut, tubing size, and produced GORs, and thus have different production potential. This is reflected in the gas lift response curve of the wells. In order to maximize oil production, the limited supply of lift gas may be distributed using the equal percentage slope method, achieving the same incremental slope in all the wells. This approach takes into consideration the production potential of every well.

The goal is to maximize the incremental return per unit of lift gas. The overall revenue is the only criterion in deciding the distribution of the extra pockets of lift gas among wells.

Fig. 1-9 illustrates how gas allocation is done based on the equal percentage slope method in a field that has five wells with different production capacities and different gas lift response curves. Note that the curves show the oil production rate vs. the lift gas rates. The assumption is that the produced water treatment costs are not significant. All wells, whether big- or small-volume oil producers, are operated at the same incremental slope.

While this method handles the limited supply of lift gas, it does not consider the effect of pressure drop in the production and gas lift lines of the field.

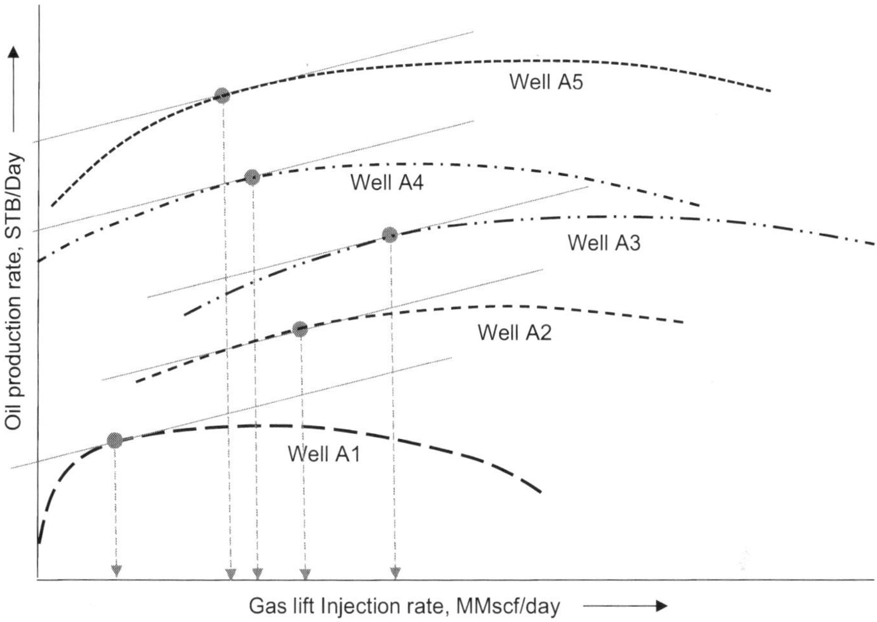

Fig. 1–9. Gas allocation by equal percentage slope

1.2.2 Lift Gas Allocation during a Maintenance Activity

The scenarios of gas lift *shortage* and *excess* should both be discussed:

- Gas lift supply in shortage:

  When there is a shortage of lift gas supply (e.g., a compressor is shut down for maintenance), the available lift gas should be redistributed. In a mature field, if there is a shortage of 5–10% of lift gas, the available gas may be redistributed to keep all the wells online. But if the shortage is huge (say, 25% or more), then some wells may have to be shut in to reduce the gas lift demand during the compressor maintenance period.

- Gas lift supply in excess:

  If a production platform is shut in for a planned activity (rig move, pipeline pigging, etc.), lift gas will be available in excess during this time. A decision should be made to shut down (or speed down or partially unload) one of the compressors—or increase the lift gas rates to the wells that are flowing during this period.

In either case, the option that maximizes the production gain should be chosen. In addition to the production gain, logistic and operational issues would also need to be considered. The level of automation in the field plays a vital role in these decisions.

In order to handle a gas shortage due to compressor maintenance, the option of shutting in a few wells is weighed against reducing a small amount of lift gas to many wells. Factors such as duration of downtime, size of the network, and number of wells affected play an important role in this decision. This is discussed in more detail in section **1.3.3** on *gas lift strategy.*

1.2.3 Use of Gas Lift Injection Hit Lists

From an oil production perspective, the well with the least efficient gas lift well has the highest injection gas-oil ratio (IGOR). These are the wells that require large quantities of lift gas to produce small volumes of oil. At times of gas lift shortage due to shutdown of gas lift compressors, wells with the least gas lift efficiency would be shut in to minimize the loss of oil production.

A well ranking in *decreasing order of IGOR,* called a *gas lift injection hit list,* is prepared and kept ready for use of operations. Starting from the top of the hit list, wells will be shut in during a compressor outage to reduce the demand of lift gas. If there is any well that should not be shut in due to reservoir management reasons, the well would be excluded from the gas lift hit list.

Even though the hydraulic efficiency (liquid rate produced/lift gas injected) is the main focus during normal gas lift operations, a well efficiency based on oil production rate/lift gas injected is used during periods of short-term gas lift shortages. This is the reciprocal of IGOR and is usually expressed in BO/MMscf.

The use of a gas lift hit list is demonstrated with a field example.

Table 1-2 shows the recent well test data of the 18 wells of the FUTURE Field from five offshore platforms. There are two gas lift compressors working in parallel, each with a capacity of 30 MMscf/D. With all the wells flowing, the field produces 30,800 BOPD, with a total liquid production of 66,000 BLPD against a total gas lift rate of 60 MMscf/Day.

One of the compressors should be shut down for two days for a planned maintenance, and the field should be managed with one compressor running during this time.

Table 1–2. Production test data for all wells

Well No	BLPD	BS&W (%)	Lift gas, Mscf/D
A-01	3562	70	4000
A-04	5004	25	4000
A-08	652	0.2	1000
A-09	4500	45	5200
A-45	1465	23	2000
B-01	6500	65	5000

B-06	2424	40	2100
B-07	6100	60	5500
B-12	2500	18	2000
B-15	4200	42	3900
C-02	2020	9	1900
C-06	980	50	1200
D-06	1750	6	1600
H-01	9800	92	8500
H-06	8600	88	7500
H-21	780	1.5	1000
H-22	2200	16	1600
H-23	3000	10	1950
	66037		**59950**

The gas lift injection hit list prepared based on well efficiency ranking for the field is shown in **Table 1-3**. The list of wells is rearranged such that the well that is least efficient in the use of lift gas is on the top of the list.

The list will be updated periodically with the latest production test data of the wells. This ensures that the field has an established plan to manage short-term gas lift shortages with minimal loss of production.

Table 1–3. Gas lift hit list

Well No	BLPD	Cum liq, BLPD	BS&W %	Lift gas, Mscf/D	BOPD	IGOR, scf/BBL	GL efficiency BOPD/ MMscfD	Cum gas, Mscf/D	Cum oil, BOPD
H-01	9800	9800	92	8500	784	10842	92	8500	784
H-06	8600	18400	88	7500	1032	7267	138	16000	1816
A-01	3562	21962	70	4000	1069	3743	267	20000	2885
C-06	980	22942	50	1200	490	2449	408	21200	3375
B-07	6100	29042	60	5500	2440	2254	444	26700	5815
B-01	6500	35542	65	5000	2275	2198	455	31700	8090
A-09	4500	40042	45	5200	2475	2101	476	36900	10565
A-45	1465	41507	23	2000	1128	1773	564	38900	11693
B-15	4200	45707	42	3900	2436	1601	625	42800	14129
A-08	652	46359	0.2	1000	651	1537	651	43800	14779
B-06	2424	48783	40	2100	1454	1444	693	45900	16234
H-21	780	49563	1.5	1000	768	1302	768	46900	17002

Well No	BLPD	Cum liq, BLPD	BS&W %	Lift gas, Mscf/D	BOPD	IGOR, scf/BBL	GL efficiency BOPD/ MMscfD	Cum gas, Mscf/D	Cum oil, BOPD
A-04	5004	54567	25	4000	3753	1066	938	50900	20755
C-02	2020	56587	9	1900	1838	1034	967	52800	22593
B-12	2500	59087	18	2000	2050	976	1025	54800	24643
D-06	1750	60837	6	1600	1645	973	1028	56400	26288
H-22	2200	63037	16	1600	1848	866	1155	58000	28136
H-23	3000	66037	10	1950	2700	722	1385	59950	30836
Total	**66037**			**59950**	**30836**				

The cumulative oil lost and cumulative liquid lost may be plotted against cumulative lift gas rate. Such a plot is very useful for larger fields with hundreds of wells and many compressors. The overall fluid and oil production loss for various lift gas shortage scenarios can be quickly estimated from this plot (**Fig. 1-10**).

In this example, for a loss of 50% of injection gas supply, the total fluid production lost is about 55%, but the oil production lost is only 24%.

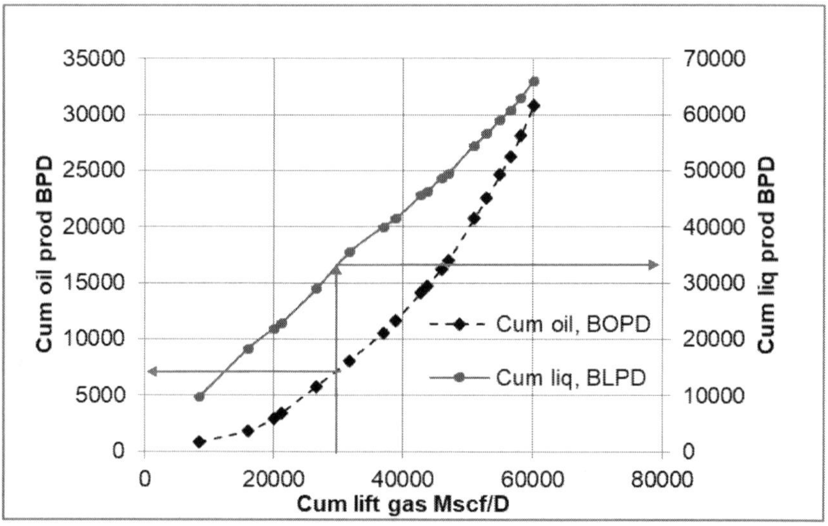

Fig. 1–10. Gas lift hit list cumulative plots

In case of an unexpected shutdown ("trip") of a compressor, it would be necessary to reduce the gas lift consumption quickly, in order to maintain a healthy gas lift manifold pressure in the system.

1.2.4 Moving beyond Well Gas Lift Optimization

Nodal analysis is a powerful tool for well modeling, analysis, and optimization. It is ideally suited for a single well analysis. While performing a sensitivity study with nodal analysis, other parameters are treated as constants.

For example, while generating a gas lift response curve, the gas lift rate is varied, but the analysis assumes a constant wellhead pressure and ability to inject any amount of lift gas. In reality, this would not be true. As the gas lift rate is increased, the flowing wellhead pressure would increase, and the additional back pressure would affect the fluid production gain from the well. Moreover, lift gas supply pressure and volume are limited by the compression capacity existing in the field.

A single-well gas lift optimization does not consider the interaction among wells in a network. In a closed rotative gas lift system, some of the key operating parameters in the field such as compression horsepower, compressor discharge pressure, compressor throughput, and separator pressure can be varied within limits. Therefore, the parameters such as gas lift injection rate, gas lift injection pressure, and flowing tubing head pressure used for predicting oil production in nodal analysis are not fixed values. The optimization program should be able to treat these parameters as *dynamic variables.*

Fig. 1-11 shows a simplified schematic of a closed rotative gas lift system. The interaction between the production (LP side) and gas lift (HP side) of the network can be understood from this figure. The two sides are interdependent—any change done to one side will affect the other.

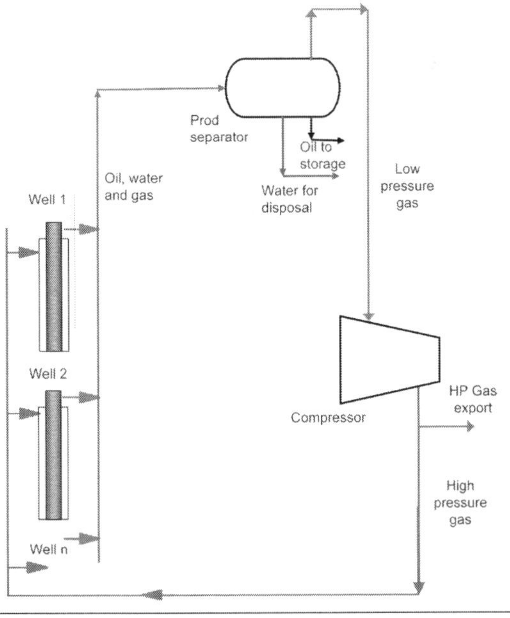

Fig. 1–11. Closed rotative gas lift system

For an operating field, what is needed is a system optimization model that would *determine* the optimum lift gas rate, lift gas pressure, and separator pressure for the current operating conditions. The lift gas quantity so determined should be within the installed compression capacity and distributed in the most efficient manner, maximizing field production. The pressure drop in the gas lift distribution and oil production pipelines should be within acceptable limits.

A holistic and simultaneous optimization of the asset would not only consider the revenue of oil and gas production, but also take into account the cost of produced water treatment, equipment maintenance, and power consumed. This would maximize the production efficiency of the field. In order to achieve a near-perfect solution, the whole field network—gas compression, gas distribution, well performance, production flowlines, trunk lines, production separator, fluid pumping, and water treatment—should be considered together. In other words, this is achieved by modeling the full network and not just the wells.

Apart from oil and gas production, the optimization model should be able to honor and implement certain non-negotiable objectives. Some of these may be field-specific, such as the following:

- A gas export should be maintained at a minimum rate of, say, 20 MMscf/D at a specified minimum pressure, at all times.
- The optimization model should be able to limit the amount of impurities at the delivery point. This may be a contractual requirement. For example, CO_2 in export gas should not exceed 2%; H_2S content of the export gas should not exceed 5000 ppm, etc.
- The optimization system should be able to impose and manage temporary constraints such as the following:
  - Only two of three main oil pumps are available (third pump is under maintenance)
  - A specific pipeline is under maintenance (pigging operations in progress, so the flow will be rerouted through another line)
  - A pipeline has been de-rated due to corrosion conditions, and pressure should not be allowed to exceed a certain limit

All of these should be achieved without compromising the main objective of maximizing the profits, i.e., increased revenue and reduced expenses.

These considerations are beyond the realm of nodal analysis. A more versatile and holistic system optimization model is needed for the full field optimization solution.

1.2.5 Optimum Lift Gas for a Field

As wells start producing water, their gas lift requirement increases. In any field, there will be some sort of in-field development or remedial activity being pursued to control the increase of water cut. As water cut increases, the company should examine the current field operating slope and decide how much additional lift gas capacity they want to build. Many wells produce at high water cuts for long periods of time, especially if the reservoir pressure is maintained. So a decision should be made to shut in wells when they reach a certain level of water cut, and this cut-off limit should be determined based on the available compression capacity of the field.

Increasing the compression capacity to keep wells with very high water cuts on production should be based on detailed economic evaluation of the scenario. Usually the economics is not favorable, as the gas compression and water treatment costs override the benefits of increased oil production. For example, a well may require about 5.0 MMscf/Day to produce 5000 BLPD at 98% water cut. Assuming a lift gas compression cost of $1000/MMscf, water disposal cost of $0.5/BBL, and an oil price of $30/BBL, the economics are clearly negative, as the operator has to spend about $7500 to make revenue of $3000. However, at water cuts less than 95%, gas lifting the well becomes profitable. These investigations can be performed accurately with the help of a full network model of the field.

1.3 Network Optimization and Gas Lift Strategy

1.3.1 Scope of Network Optimization

The terms *system optimization* and *network optimization* are used synonymously in the current context.

As a broad concept, to *optimize* means to plan or carry out an economic activity with maximum efficiency, taking full advantage of the flexibilities of the existing process and systems. Optimization would also take into account the several, often conflicting, requirements in an operating environment and arrive at the best compromise that maximizes value.

There are two types of field optimization opportunities that arise with any asset:

- **Field life optimization**

  Long-term planning to help determine the facilities that are required to get the best output from the field. The goal is to maximize *net present value* (NPV) or *rate of return* (ROR), all the time adhering to the corporate rules as laid down by the government or shareholders.

- **Snapshot optimization**

  Short term planning to efficiently utilize the existing facilities. This entails maximizing a value function—cash flow—and is subject to constraints imposed by other factors such as existing well deliverability, facilities capacity, contract requirements, environmental issues, and others.

The goal is to design a powerful network system optimization model that can be used to achieve the long-term and short-term objectives through all phases of the field life.

Fig. 1-12 shows the typical development life cycle of a field and the system optimization requirement during different phases of the field life. [3]

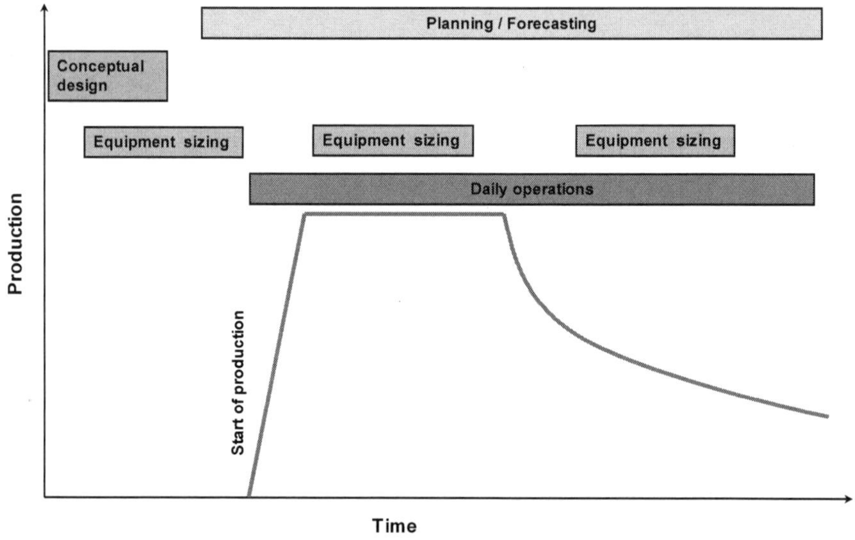

Fig. 1–12. System optimization in all phases of the field

The following are the major steps of field life that will require system optimization:

- Conceptual design
- Equipment sizing before commencement of production and as the field matures
- Daily operations till the field is abandoned
- Planning and forecasting till the abandonment of the field

1.3.2 Importance of Considering Full Network for Optimization

In a gas lifted field, the pressure drop in the production system and gas lift distribution system affects the production from the field.

Pressure drop in production side:

Offshore, if the wellhead platform is far away from the central facilities platform, depending on the pipeline size and flow rates, there could be a significant pressure difference between the production manifold at the platform and the production separator at facilities. This would result in higher flowing wellhead pressures. Similar pressure may exist in onshore assets too, if the well location is far away from the separation area.

If more wells are added in the field over a period of time due to in-fill drilling, the production manifold pressure will gradually increase due to increased fluid load and congestion in the production gathering network. The production separator is usually operated at the same pressure.

In both these cases, the gas lifted wells are forced to flow against high wellhead flowing pressure. This will make the wells less efficient.

Pressure drop in gas lift side:

While the gas compressor performance may be normal, the gas lift header pressure at the platform may be lower than design, due to excessive pressure loss in the gas pipeline from compressor location to the wellhead platform. Length of the pipeline, restrictive ID, liquid loading, and gas flow rate exceeding the design may lead to this situation. This will result in shallow gas lift injection in wells.

The gas lift supply pressure may also be affected by compressor performance issues occasionally. This may have some impact on some wells but only for short durations.

In large fields, multiple compressors work in parallel to feed high-pressure gas into a common gas supply manifold. The gas lift rate to individual wells may be controlled by flow control valves working as gas lift chokes. If automated, these chokes maintain the set point gas lift rates to the wells. If set points are increased, more gas flow will be drawn from the compressors, and the compressors would work harder to maintain the discharge pressure with the increased throughput.

Fig. 1-13 shows the detailed schematic of a rotative gas lifted network. Also shown in the figure are some of the major parameters that can be varied to achieve the best operating conditions for the asset at any given time, subject to existing field constraints. The parameter that can be optimized is either pressure (P) or flow rate (Q) as marked in the network [(3)]. The items that add to the revenue are designated as +$ and the ones that add to costs shown as -$.

As a well is connected to both gas lift and production sides of the network, anything that happens to the network can affect the individual well performance. The reverse is also true—a change that happens in a well can affect the network too. For example, if a new gas lifted well is commissioned on a platform, depending on the sizes of the pipelines and flow rates, the gas lift manifold pressure may decrease and the production manifold pressure may increase. This can happen even if the gas compressor discharge pressure and the production separator pressure remain unchanged. This will affect the performance of other wells connected to the network. This is due to the fact that the injected lift gas is returned to the production system at low pressures along with produced fluids.

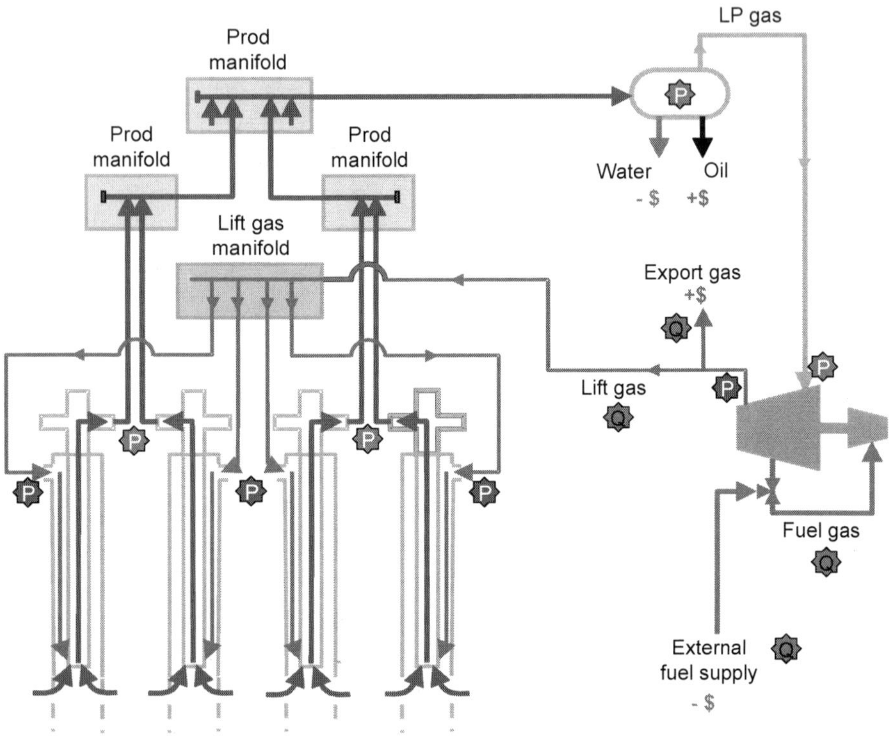

Fig. 1–13. Rotative gas lift network with optimization parameters

Optimizing a well in a stand-alone manner would not be a true optimization, as the well is a part of a bigger network. Therefore, the objective should be to optimize the full network. This would lead to different possibilities in field operations, i.e., realizing the full potential of the existing equipment and managing the existing constraints. This, in turn, would improve the business results of overall field operation. In general, inclusion of the gas compression and gas injection system enhances the overall merit of the optimization solution, in comparison to the production pipeline model alone. [4]

As an example, the impact of varying the production separator pressure in a gas lifted field may be reviewed. As the separator pressure is controlled in the field, increasing the separator pressure would lead to an increase of the production manifold pressure. If the well production is not controlled by choke, the flowing wellhead pressure would increase, leading to a decrease in production rate from the wells. On the other hand, increase of the separator pressure increases suction pressure of the gas compressor. If the compressor discharge pressure is maintained constant, increase of suction pressure will decrease the compression ratio, and this will reduce the horsepower requirement for the same quantity of gas compressed. The overall impact is that some production would have been lost, and fuel gas requirement for compression would have decreased. At times of shortage of compression horsepower, this approach helps to maintain the gas lift supply pressure of the network.

The solution for a network optimization would be iterative, considering the capabilities and constraints of the total system. A system optimization should consider:

- Well performance for each well, taking into account its current inflow and outflow potential
- Effect of pressure drop in the production surface network system
- Compression horsepower, performance, availability, and fuel or power costs
- Effect of pressure drop in the gas lift supply system
- Alternate sources available to the field in terms of the supply of fuel and high-pressure gas

Taking a holistic view of the system availability, current efficiency of wells, and condition of equipment and pipelines, a network optimization model will generate optimum values for the following parameters to be used in the field:

- Production separator pressure
- Gas lift header pressure
- Gas lift injection rates for individual wells
- Gas lift compression horsepower
- Use of additional resources available to the field

This can be done for normal operation of the field, as well as during any maintenance activity—maintenance of compressors, pumps, separators, heat exchangers, and critical pipelines.

1.3.3 Gas Lift Strategy

If gas lift is the major source of production for an operating company, the organization frames a gas lift strategy that would maximize production at all times.

1.3.3.1 Gas Lift Strategy for Single Field Operations

A strategy document explains the production impact of any operation likely to happen in the field, and suggests ways of minimizing loss of production. It covers various operational scenarios of short-term and long-term gas lift shortages arising out of planned and unplanned operations. It serves as a tool of communication to all the parties concerned so that they are aware of the production impact of the various options available. The strategy document is jointly generated by engineering and operations and updated at regular intervals. Input from specialized personnel such as equipment and process engineers is incorporated into the strategy so as to ensure smooth operation and long life of the associated equipment.

Referring back to the schematic of the detailed gas lifted network (**Fig. 1-13**), the network model can be used to evaluate the impact of different operational scenarios in large gas lift networks, such as the following:

- Expected increase of oil production for a small decrease in separator pressure (say, 5 psi) and corresponding increase of horsepower needed to maintain the required gas lift supply
- Impact of reducing the gas lift supply pressure (say, by 100 psi)
- Impact on oil production, gas lift demand, and produced water treatment, if wells with high water cuts (say, more than 97%) are shut in in the field
- Optimum quantity of high-pressure gas that should be imported when one of the gas compressors will be shut down for three to four days for an engine change

A system optimization model will be of great help while developing a gas lift strategy. Many optimization investigations would be performed with the system optimization model to frame and fine-tune the strategy. The strategy provides the means to maximize the production efficiency of the asset and meet production deadlines. Operationally, the strategy will aim to maintain a steady gas lift supply pressure without allowing the pressure to fall below a minimum value. The use of a gas lift hit list to shut in wells for handling a temporary gas lift shortage is a part of the strategy.

The strategy document addresses both situations: available gas lift compression being in shortage or excess. The outline of strategy recommendations is indicated below.

- If gas compression capacity is in shortage, the system optimization model will be used to investigate the impact of the following actions:
    - Adjust speed of the running compressors / optimize gas discharge pressure versus throughput
    - Reset the gas lift rate set points, decreasing lift gas to a few wells, where possible
    - Increase the separator pressure
    - Shut in wells, as needed, to maintain a minimum gas lift header pressure
- If gas compression capacity is in excess, the system optimization model will be used to investigate the impact of the following actions:
    - Bring all wells online
    - Reduce separator pressure
    - Adjust compressor speed / optimize gas discharge pressure versus throughput
    - Reset gas lift rate set points, increasing lift gas to a few wells, as needed
    - Keep one of the compressors on stand-by mode, if possible

1.3.3.2 Gas Lift Strategy for Multiple Field Operations

A company may operate more than one field. Each field may have different levels of gas lift supply and demand. Each field has its own process flexibilities and limitations. At times of need, these flexibilities can be exploited fruitfully, and resources can be shared to achieve corporate goals. The gas lift strategy aims to bring out and nurture synergies between the fields.

The strategy ensures operations make best use of the available resources between the fields, during normal operations as well as maintenance activities.

- Under normal operating conditions, one field may have more gas compression capacity than the other. It would make sense to export some high pressure gas to the field that needs it, within operational limits.
- If one of the gas lift compressors is taken down for maintenance in one field, the gas transfer can be adjusted to help the field where the compressor is down. This will help minimize overall production loss. However, the volume of gas lift gas taken away from a loop cannot exceed the associated gas produced by the field.

The gas lift strategy is an outcome of teamwork within an organization. Production engineering and operations will work with maintenance teams to frame and follow the strategy. It will be a living document, updated by the optimization team at regular intervals as field conditions change.

2

Network Optimization Becoming a Reality

Having established the benefits of a network optimization approach, it is necessary to develop a fundamental understanding of how to achieve it in a practical sense. With the field being very complex, it is necessary to draw some lines as to what should be included in the model. Given the limitations of time and money, decisions should be made on the selection of modeling technology that would be best suited for the task at hand. The entire process should be thought through, starting from data collection for building the model to permanently maintaining a calibrated model as a working tool. The company should motivate itself to utilize the network optimization model in various ways and enjoy technical and commercial benefits.

This chapter presents a discussion of the following topics:

- Model-based optimization approach
- System optimization growth steps
- Components of a network model
- Selection of the optimization engine
- Functional requirements of a system optimization model
- Practical aspects of system optimization implementation
- Various benefits of the system optimization model

2.1 Foundation for System Optimization

2.1.1 System Optimization—Pioneers

Even though the idea of network optimization has been around for many years, the practical network optimization of large gas lifted fields with a complex rotative gas lift system with loops and branches was first implemented by a team of professionals from Edinburgh Petroleum Services Ltd. (EPS Ltd.), Edinburgh, UK, that later became part of Weatherford Production Optimization. This was first implemented in a giant field in the Middle East with more than 250 wells. The network had many pumps, compressors, and separators operating in a parallel manner.

Along with the software, the company developed a three-step workflow process called *Aim-Gain-Sustain*™ and delivered successful network optimization solutions for many operators.

AIM:
The first step of the process is an Asset Inspection Methodology (AIM) that defines feasibility, business case, and costing.
GAIN:
The next step is engineering consulting work that gives a fast payback.
SUSTAIN:
The third step is to implement the technology, transform processes, and train people so that the operator can establish a sustainable production optimization system in the asset. This would be eventually maintained by the staff of the operating company.

This business model has a few key differences compared to the work done by traditional service companies. The process provides early value to the stakeholders, which pays for the implementation along the way. The biggest attraction of the process is that it integrates technology, processes, and people, as shown in **Fig. 2-1**. The workflows in the organization would be modified to accommodate the new optimization processes. People from the operating company would participate and get trained as a part of the project. This allows the operating company to take ownership of the optimization process and enjoy benefits for the long term.

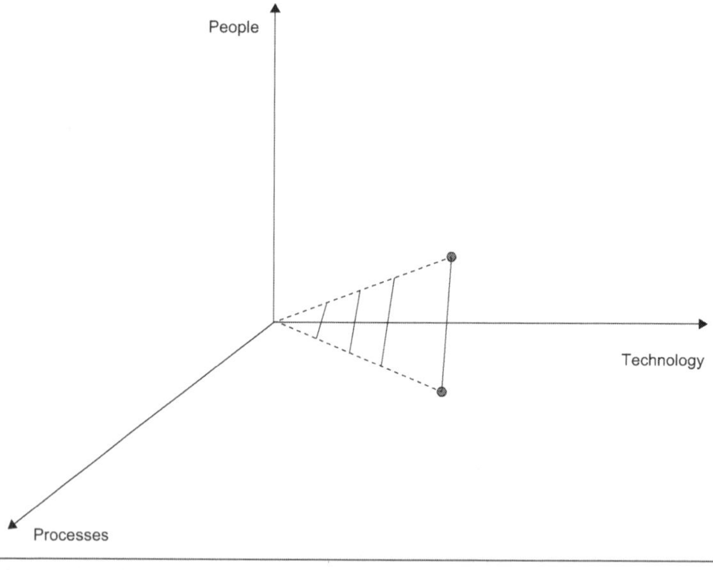

Fig. 2–1. Integration of people, technology, and processes

Other companies who subsequently developed similar optimization solutions for the industry have also been very successful.

2.1.2 Model-based Optimization

The optimization using physics-based models makes the best use of the engineering principles behind the functioning of the various pieces of equipment of the asset.

There are two aspects that are honored well in a model-based optimization approach. First, the individual pieces of equipment are modeled accurately. The model goes to the roots of the design and incorporates the performance characteristics of every component, including the performance boundaries of the devices. The optimization model will use these details in a dynamic situation.

Second, the connection between various components is properly established in the model, as found in the field. In the case of pipelines, the details would include the size (ID and OD), length, and material of construction of the pipes, along with insulation details. This ensures accurate calculation of loss of pressure and temperature in pipelines by the model.

For example, a wellhead choke is modeled using a choke equation applicable for the service. When the choke opening is increased, the flowing wellhead pressure will decrease and well production rate will increase. The new production rate will be based on the well performance curve of the subject well in the model. The choke downstream pressure will stabilize based on the nature of flow through the choke (critical or non-critical) and the pressure drop in the pipeline leading to the separator.

All components of the field should be modeled using their engineering characteristics. Only then, the network model will be able to utilize the flexibility associated with the network components and achieve superior results during optimization. For example, if a pump is modeled with its design performance curves, for a fluid rate pumped, the model predicts the head developed and calculates the horsepower required. When the speed is increased, the model follows a different performance curve and performs similar calculations for the changed conditions.

If this approach is used, the physics-based model will utilize the characteristics of the object in the model and process the interactions between the objects accurately. However, it should be noted that, in the offline mode, the model is stand-alone—not connected to live data. So if the performance of a compressor is deteriorating, the model will not know this unless someone recalibrates the compressor performance curve in the model. If the head developed is lesser than what is predicted by the original compressor performance curve, the head curve should be adjusted in the model during the recalibration, based on actual field data, to capture the performance deterioration.

2.1.3 System Optimization Growth Steps

The growth grid for gas lift system optimization in most of the assets usually follows four steps in the sequence shown in **Fig. 2-2**.

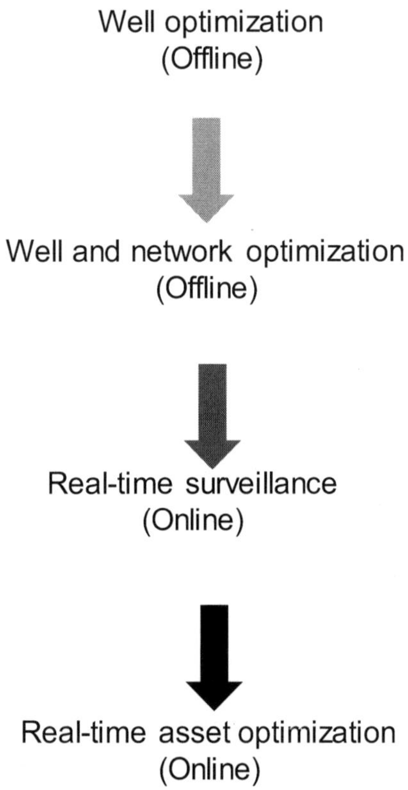

Fig. 2–2. Growth grid for gas lift optimization

Step 1: Well gas lift optimization (Offline)

Well models are built using reservoir and completion data and calibrated using operation data from the field. The models are used offline. The word "offline" implies that the models are not directly linked with field operation, but the models are used in a parallel process. Frequently (say, once a month), the data from the field wells is used to calibrate and update the well models. Optimization runs are performed on the well models and the set points adjusted in the field. Production engineers / gas lift specialists and production operations are involved in the process.

In an effort to minimize the time spent on manual calibration of well models, engineers try to calibrate the well models in a semi-automated way by transferring the production test data directly from the well test database into the model and running a calibration routine.

Step 2: Well and network gas lift optimization (Offline)

This is a major step. A network model is built. The updated well models from Step 1 are incorporated in the network model. In addition to the wells, the network model has all the essential process components that are part of the field gas lift operations:

- On the production side, it has the details of wellhead chokes, flowlines, and production gathering systems for all wells, up to production separator.
- On the gas lift side, it has the details of the LP gas pipelines from the production separator to the compressors, the compressor models, and HP gas distribution pipeline network from the compressor to the wells or export line.
- Pipeline details include internal diameter (ID), length, insulation details, etc.
- Pump and compressor data includes performance curves, horsepower details, and other data.

The status of wells, valves, chokes, and other pieces of equipment is set in the model as it exists in the field. The full network model is history-matched using operating data obtained from the field. The model after history match is analyzed to identify any well, pipeline, or any other component that displays any abnormal behavior. Remedial actions are initiated to improve the performance of these components.

Optimization runs are performed by running the calibrated network model with specific optimization objectives. Optimum values of gas lift rate, gas lift pressure, and separator pressure obtained from these model runs are reviewed and implemented in the field. As the system stabilizes, results are tracked and gains monitored. The well and network models are re-calibrated with the new stabilized data collected from the field and the optimization cycle repeated.

This process is repeated as frequently as possible depending on the size of the field. The models are still offline.

Step 3: Establish real-time surveillance

Sensors are installed at critical points of the asset and connected to a data-gathering system. Once the data is collected in the digital format, the data can be processed or grouped in several meaningful ways. This can be displayed in digital format or graphical format on a real-time basis. The visualization can be displayed on any computer screen connected to the server, irrespective of the physical location. This means that the current data of the asset and the performance trend can be displayed at many locations in real-time—production office, technical services, company headquarters, and so on. This is called *real-time data surveillance.*

With implementation of this system, the access to data becomes easy. All authorized personnel can review well or equipment status irrespective of their physical location. They can see the data or the trend plots on their smart phones if they are on the move. The good thing is that all will see only *one version* of the information at a given time.

Retrieving digital data from facilities in old fields is relatively easier, as the control system of the facilities would already be collecting the data in some form. Collecting data from wells and pipelines in old fields is a challenge if sensors are not in place or data communication facilities do not exist. The data should be collected and transmitted through cables or wireless methods. Until this is accomplished, the well data from the field may be manually input into the system. Manual data collection is tedious and will require many person-hours. It may not be practically possible to execute the manual data entry on a frequent basis in large assets. Setting up the required instrumentation is the permanent solution.

In most of the new fields, the data capture and real-time transmission systems are pre-built.

Step 4: Real-time optimization

This is the final step in the process. The total system optimization cycle using offline models is already working. The real-time data is available in digital form. Keeping the models online, the real-time data is fed to the models. The models are calibrated online. Optimization runs can be performed with the calibrated models online. The validated results of the optimization runs can be directly sent to the field to implement as new set points. This is called *real-time optimization* (RTO).

An intelligent supervisory system should be put in place to manage the RTO process:

- Perform QA/QC of the data
- Route the good-quality data to the respective models
- Calibrate the models and send out an alert message about the result of the calibration
- Run the optimization routine and communicate the results of system optimization to the assigned teams for QA/QC or for direct field implementation

The RTO system should be built and customized to meet the business needs of the asset. The frequency of optimization is decided based on the business requirement. This could be once a day or on an instantaneous trigger, depending on the business need.

The RTO is designed to assist the site management in performing the asset functions in a very fast and reliable manner. Matured RTO systems generate recommendations on a daily basis as to how to maximize the value of the asset while meeting contractual obligations imposed at the corporate level.

2.1.4 New Basis of System Optimization

There is a revenue and cost associated with any petroleum production operation. In gas lift operations, sale of produced oil and gas constitutes the revenue from the field. Water production, power or fuel, and equipment maintenance charges make up the costs. The cost for lift gas is actually the cost of gas compression as the gas is returned at low pressure along with produced fluids. Produced water treatment is required to make it suitable for reinjection or disposal.

$$\text{Profit} = \text{Revenue} - \text{Cost}$$

The total system optimization should take all the revenues and costs and aim at maximizing the net profit. Thus, the new basis of network optimization is the *net profit* ($/Day) and not BOPD alone.

A simple example of two gas lifted wells in a network will be considered to illustrate the profitability of gas lift operations. The impact of back-pressure is ignored for this discussion.

Well A: Liquid production rate 4000 BPD, water cut 10%, using lift gas rate of 2.5 MMscf/D.

Well B: Liquid production rate 4000 BPD, water cut 90%, using lift gas rate of 5.0 MMscf/D.

Using an oil price of $30/BBL and water processing cost of $1.0/BBL, and with cost of gas lift gas being charged at $1000/MMscf, and neglecting the revenue from the associated gas produced, it is clear that operating well B under these conditions makes marginal profit.

Well A: 3600*30 – 400*1.0 – 2.5*1000 = +$105,100/Day

Well B: 400*30 – 3600*1.0 – 5.0*1000 = +$3400/Day

Holding all other values unchanged, if the oil price increases to $40/BBL, the profit from well B improves slightly, although it is much less than well A.

Well A: 3600*40 – 400*1.0 – 2.5*1000 = +$141,100/Day

Well B: 400*40 – 3600*1.0 – 5.0*1000 = + $7,400/Day

It follows that the price of oil and the cost of lift gas and produced water treatment controls the economics of day-to-day gas lift operations. If the water cut of well B increases to 95%, even with an oil price of $40/BBL, the profit is negative. Operations should consider alternate use of the lift gas to produce more oil from other wells.

Well B: 200*40 – 3800*1.0 – 5.0*1000 =–$800/Day

Even though certain safety and environmental issues that do not make "profit" are very important, the profitability of well operations cannot be ignored.

2.1.5 Components of a Network Model

A network modeling software is needed to build a network model. The goal is to construct an engineering model that will replicate the field for the purpose of simulating the field operations.

The model can simulate the effect of fluid flow in a network and predict the pressures and temperatures along with flow rates at various locations. In many aspects, a model cannot represent the dynamics of the asset. For example, safety and integrity of the asset cannot be predicted with the model.

The model is built to reflect steady-state or semi-steady-state flow conditions, and not transient conditions. In other words, only stabilized operations are modeled, not field start-up or shut-down. There are other tools available in the market to perform transient simulations.

Fig. 2-3 shows the operation schematic of an oilfield. It is necessary to include all of the components that can affect production in a network model. It is also necessary to define and include the performance characteristics of all of these components. The objective being production optimization and not process optimization, the components that are included in the model and their level of detail should be sufficient to achieve the optimization objective. It is also dictated by the level of accuracy or resolution of the solution that is expected from the model.

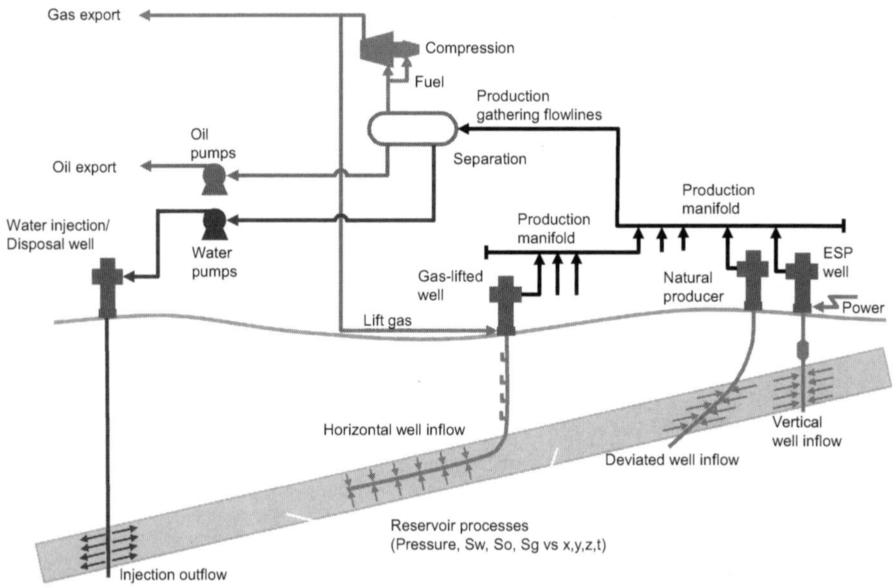

Fig. 2–3. Oilfield operation schematic

The objects that should be included in gas lift network models are listed in **Table 2-1**:

Table 2–1. Category objects in network model

S. No.	Object to be modeled	Example details
1	Wells	1. Producer wells – Natural flow oil producer, Gas lift oil producer, ESP, or other pumped wells 2. Injector wells – Water injector, Gas injector, WAG injector (Water alternating gas injection) 3. Water wells – Water supply well, Water disposal wells
2	Pipes and pipe fittings	1. Flowlines and pipelines, headers, trunk lines and oil export pipelines, LP gas lines, HP gas distribution / injection network for gas lift. 2. Pipe fittings: pipeline elbows, bends, expansion joints, constrictions, if necessary (usually modeled using equivalent length)
3	Valves and chokes	Block valves (open / close), control valves / chokes (fixed opening / variable opening)
4	Process vessels	Production separators, gas dehydration / processing, heat exchangers, other special equipment such as de-salters, LP separators, ejectors, extraction towers, etc. (as detailed as needed for the production optimization)
5	Process equipment	Pumps, compressors with set of auxiliary equipment as needed
6	Interconnection with other fields	HP gas sales (export), oil import / export, lean gas import, fuel gas / energy supply

2.1.6 Network Model Reference Depth

The network optimization model has a reference depth. It would be at surface, as a majority of the process equipment items are at this depth. Using this as the reference depth, wells and pipelines that are above or below the reference depth (overhead / buried / sub-surface pipelines) will be modeled as found in the field. The end depths will be defined in the model as + or – depths. Slant lines can be modeled as installed. The downcomers and risers can also be modeled with the actual end depths.

Fig. 2-4 shows the reference depth of a downcomer from a production platform and the riser at the separator platform offshore.

If the separator level is taken as the reference depth for the surface model, the well downcomer top end is – 15 ft and the bottom end is -190 ft. On the separator platform, the riser bottom is at -205 ft and the top at 0 ft (separator level). The model performs head and pressure calculations throughout the network accordingly.

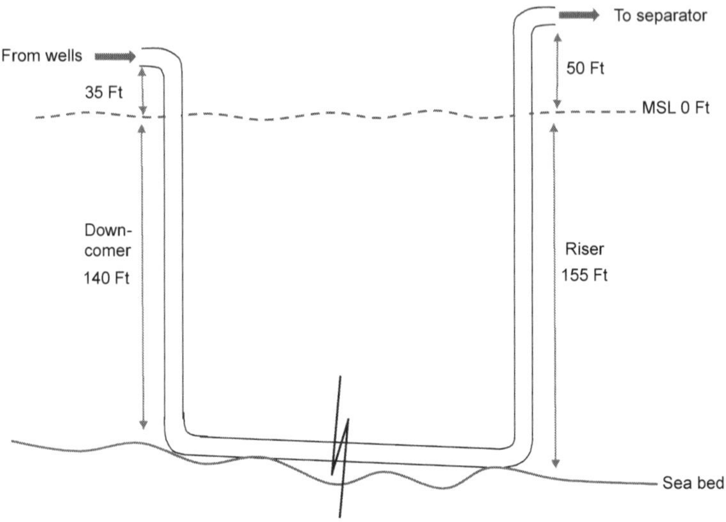

Fig. 2–4. Network model reference depth

2.2 The System Optimization Engine

2.2.1 Optimization of Non-hierarchical Networks

If fluid flows from point A to B to C in a process, the flow direction is already defined. Hierarchical network arrangement provides a pre-defined flow path.

In the production line-up shown in **Fig. 2-5**, Wells 1-4 flow into Manifold A and Wells 5 and 6 flow into Manifold B. Production from these two manifolds flows into Separator A. This represents a *hierarchical network.*

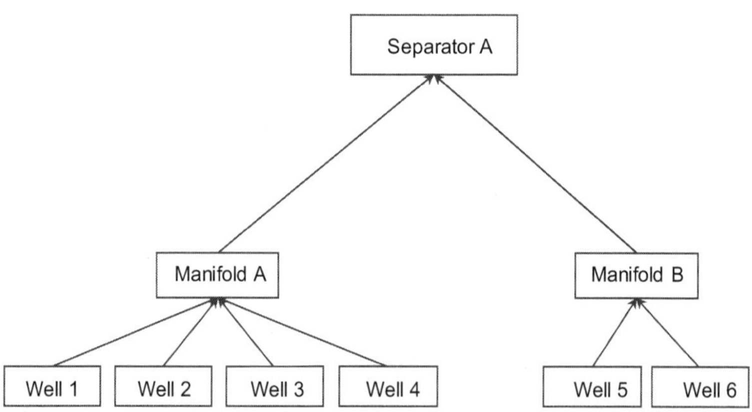

Fig. 2–5. Hierarchical production network

Unfortunately, in field production networks, the flow direction does not always follow a pre-determined direction due to the presence of parallel or alternate flow paths within the network. In the production network shown in **Fig. 2-5,** in order to distribute production evenly (to avoid overloading of either of the manifolds), headers A and B may be interconnected. Then the flow direction can be from Header A to B or the reverse. Furthermore, if a parallel separator is added to the network, as shown in **Fig. 2-6,** it is not known what percentage would go to each of Separators A and B. The network has become non-hierarchical, and nodal analysis cannot solve this problem. The nodal analysis approach can solve parts of the network individually. But here the flow in each branch is not pre-determined. In fact, the objective is to find a split in flow that will minimize the manifold pressures at Manifolds A and B, in order to maximize production from the wells.

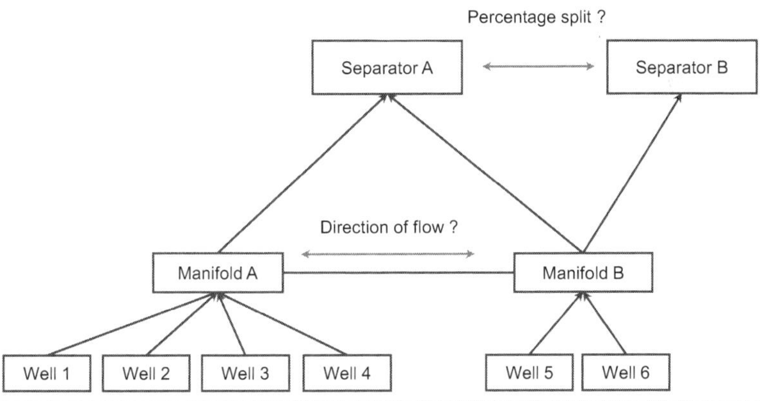

Fig. 2–6. Non-hierarchical production network

The optimizer chosen for the optimization of such complex, non-hierarchical networks with loops and branches should be:

- Capable of modeling the entire production network
- Able to simulate the production network behavior including compressors, heat exchangers, pipelines, pumps, gas lift, separators, and re-circulation
- Designed to optimize revenue, subject to the costs and operational constraints
- Able to handle the optimization required in all phases of field life, to support the associated business functions
- Able to consider the multiphase network response of multiple wells (with different PVT) producing into a common production system, where the response of one well would affect production of another
- Able to consider the network as a whole and be able to solve large non-hierarchical networks with no limits on complexity.

There are several solution options used in the industry. Two popular options are *Sequential Linear Programming* (SLP) and *Sequential Quadratic Programming* (SQP). Software solutions using these two techniques have been widely used to model and optimize large networks in the oil and gas industry.

2.2.2 Optimizer with Sequential Linear Programming

The model employing the SLP performs simultaneous simulation and optimization. It simulates or models the entire production system—reservoir inflow, well performance, surface facilities including multi-stage separation, compression, and pumps—taking into account the delivery constraints, pressures, and prices; it simultaneously optimizes production from the assets, incorporating real-world constraints such as pressure, rates, and penalties, and maximizes net revenue by balancing revenues and costs [3].

This product is called ReO™ (Resource Optimizer), developed by Edinburgh Petroleum Services Ltd., a Weatherford Company.

In SLP, the highly non-linear models describing an oil or gas field production system are linearized around a starting value. The linearized equations are optimized using a standard linear programming method. The Simplex method used is a standard technique in linear programming for solving an optimization problem, typically one involving a function and several constraints expressed as inequalities. The inequalities define a polygonal region, and the solution is typically at one of the vertices. [5]

A *trust region* is applied to limit the region in which the linearized model is "trusted" to be an acceptable approximation to the full model (**Fig. 2-7**). The trust regions improve the convergence process for the SLP solver, ensuring that the convergence behavior is stable.

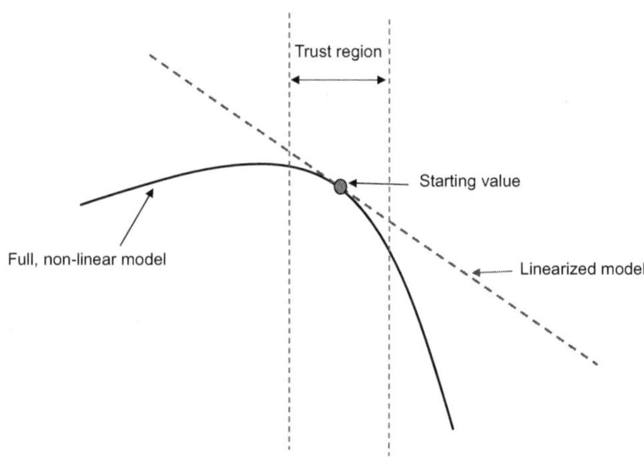

Fig. 2–7. SLP Optimizer working with a trust region

The full model equations are evaluated at the new optimum solution and compared to the linearized model. The process is repeated until the difference between the full and linearized models is less than the convergence tolerance (**Fig. 2-8**).

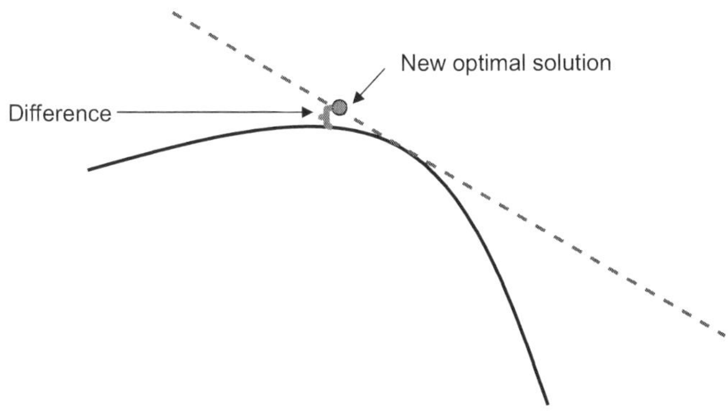

Fig. 2–8. Converged solution

This product uses advanced SLP and is able to handle non-linear responses of network elements; it is scalable and can handle any topology. Reliable convergence is obtained with accurate network element responses. Plus, it is proven in the industry in many projects involving large networks. [6]

The use of specially ordered sets enables SLP to represent non-planar performance surfaces as piecewise linear approximations. The granularity of the piecewise linear approximation can be set by the user to ensure that there is negligible difference between the true non-linear response and the piecewise linear approximation. Such a three-dimensional performance surface for a gas lifted well is shown in **Fig. 2-9**.

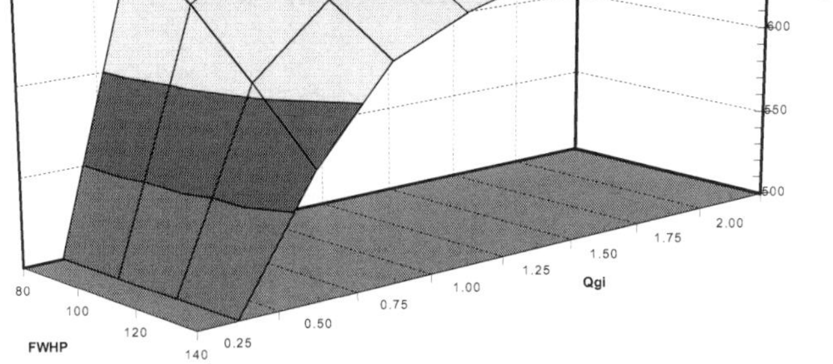

Fig. 2–9. Piecewise linear surface representing well performance

The basic logic of the solver is outlined in the following flowchart (**Fig. 2-10**).

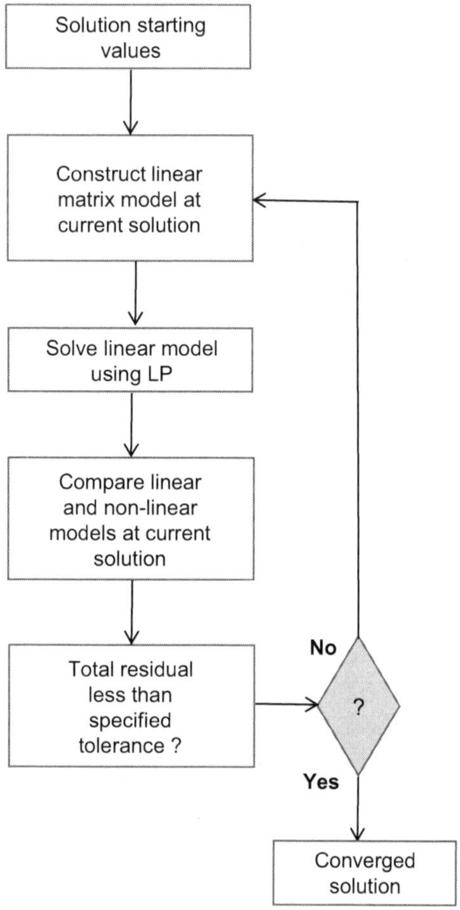

Fig. 2–10. SLP solver flow chart

2.2.3 Optimizer with Sequential Quadratic Programming

The model employing the SQP approach uses an equation-based engine for solving integrated oilfield networks to obtain optimized solutions for day-to-day operations and a rule-based network solver functionality for handling long-term forecasting by honoring constraints and evaluating long-term production goals. [7] This product is called GAP™ (General Allocation Program), developed by Petroleum Experts, Edinburgh, UK.

The core technique has two levels:

Level 1: Solving the network

Level 2: Optimizing the network response

The optimizer will achieve the maximum hydrocarbon production using the rate of change of the production rate with respect to the rate of change of controllable variables, e.g., of the injected lift gas rate for gas lifted wells.

The non-linear optimization engine of the program allocates gas for gas lifted wells or sets wellhead chokes for naturally flowing wells to maximize revenue or oil/gas production while honoring constraints at any level in the system. The program offers a fast and robust global optimization algorithm using non-linear programming (NLP).

SQP techniques work by calculating the response of every element in the system, not only on how pressures will change with changing flow rates, but also how the gas lift injection rate will impact the pressures.

The network behavior seen in a typical gas lifted field with multiphase flow may be represented to show how the function Y will vary with X. By the very nature of it, each time a guess of X is made in order to calculate the corresponding Y, the Y function itself changes, as it may be dependent on other variables affected by the choice of X. **Fig. 2-11** shows how the function Y behaves like a moving target for a choice of X:

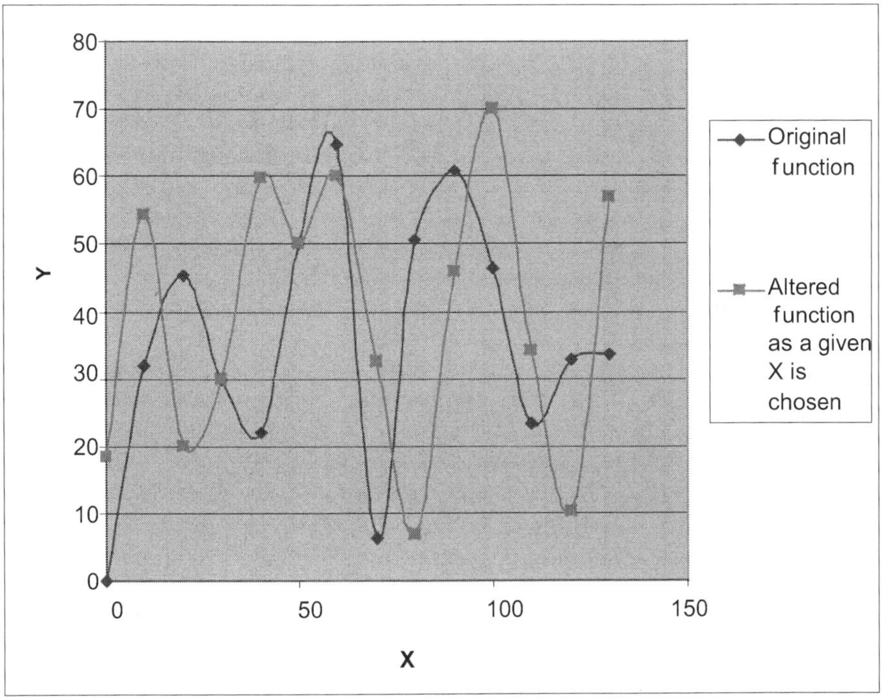

Fig. 2–11. Moving target representation

While performing lift gas allocation, the optimizer will start with the well performance curve for a fixed WHP. As a first step, a guess for injection rate is made, and the corresponding liquid rate calculated from the well response. This rate will then be used to calculate the response of the surface pipelines that will return a new wellhead pressure, based on the pressure drop in the system. A new performance curve will then be constructed for the well using the new WHP. The iterations are repeated in order to converge to the solution (**Fig. 2-12**).

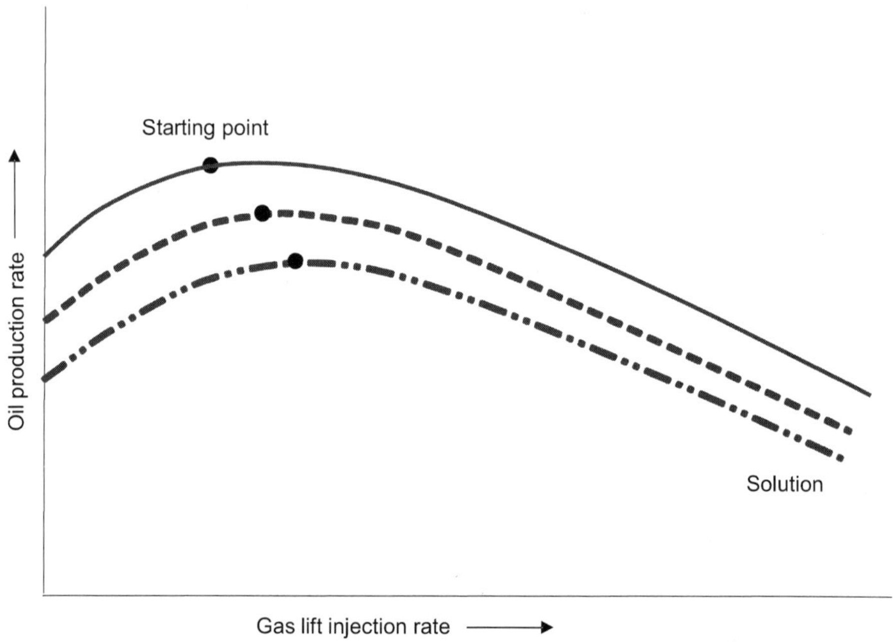

Fig. 2–12. Effect of network on well gas lift performance

The processing logic of this program for both current optimization as well as future forecasting is shown in **Fig. 2-13.**

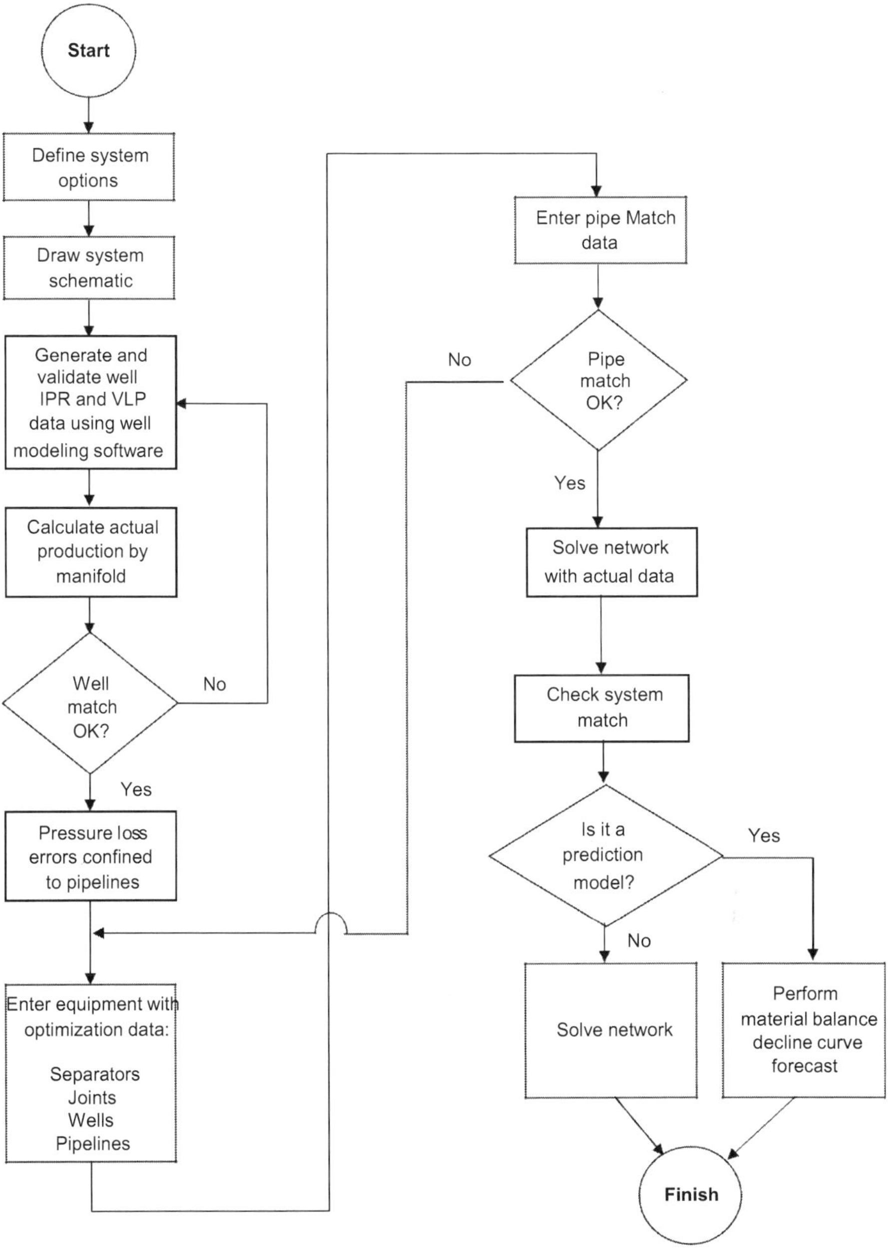

Fig. 2–13. SQP processing logic

2.3 Selection of System Optimization Model

2.3.1 Expectations on Optimization Capabilities

While screening various software products for network optimization, the operating company engineers develop an in-depth knowledge about the products by seeking to understand practical issues associated with the software. A few key aspects that should be considered before selecting the optimization software are listed in **Table 2-2**. This would help set the right expectations about the model.

Table 2–2. Network Optimization Software Expectations

Petroleum Engineering	The petroleum technology used in the model is comprehensive enough to represent the field production operations. The model can run on black oil or fully compositional PVT, based on the requirement of the asset. The model is able to handle loops and branches, and parallel compressors, pumps, separators, and pipelines.
Optimization Technology	The model is able to impose all the practical constraints that would be set up anywhere in the network. The model can be used to run optimization to deliver real-life scenarios e.g., limiting the non-organics (H_2S or CO_2) in gas exported from the network. The model has the ability to use industry-standard correlations for modeling the flow of fluids in various pipelines. The model is history-matched to the existing field conditions and optimized to reveal a better method of operation (if any). The model is stable and robust and converges with a valid solution in a reasonable time.
Usability	The product is user-friendly. It allows building large models quickly with relative ease (should use regular Windows functions to be able to copy and recreate similar areas). The graphical user interface (GUI) is powerful and easy to navigate. Should be able to link the surface network model to other essential software applications easily (e.g., well modeling, PVT modeling, reservoir modeling, etc.). The model can be populated with data in an easy manner using computer spreadsheets (i.e., without the need for manual entry). Editing of bulk data is possible as a single-click operation. The model is able to output the results easily in graphical format, if needed. The software allows extraction of input data and results in any standard format, e.g., Word, Excel, etc., in a customized manner, to compare one run with another.
Presentation & Visualization	The model is capable of showing visually which parts of the network are performing in a sub-optimal manner, e.g., high liquid holdup, erosional limits reached, unacceptably high pressure drop gradient in pipe segments, etc. The model is able to present the network in an aesthetic manner (similar to how the field is arranged) and easy to navigate for users.
Future Scope	The database required for the model can be easily linked to the other database systems of the operating company. The model can be linked to the real-time data environment at a later date to facilitate RTO. The model can be expanded when needed.

Focus of Service Provider	The software has been used in optimization of assets of similar size and complexity. The team from the service provider is willing to work closely with the operating company to achieve required success, even if their office is in a different time zone. The service provider is flexible and willing to improve their product, if and when needed. The modeling service provider has a sufficient number of specialized consultants with varied experience, who can guide and provide a clear path towards success while working on the project. The expertise required includes petroleum engineering (production technology, PVT, artificial lift), multi-phase flow, pumps and compressors, and well and facility operations, apart from production modeling and optimization.

2.3.2 Detailed Functional Requirements

The system optimization software should have the following capabilities:

1) The software tool should be capable of complete modeling of the surface production / injection network. It should handle a variety of wells in the same network:
   - Naturally flowing oil wells
   - Different types of artificial lifted wells (gas lifted or pumped)
   - Condensate or gas producers
   - Water producers
   - Water or gas injectors
2) The model should be capable of optimizing the entire production system from the reservoir to the delivery point. The network production rates may be maximized by altering certain conditions, such as:
   - Back pressure of wells by applying wellhead chokes
   - Gas-lift gas allocated to individual gas lifted wells
   - Frequency of operation of ESP-fitted wells
   - Operating speed of pumps / compressors
   - Choke / Control valve opening
3) The model should be capable of production optimization, allocation of production, and production forecast. The production forecasts could include gas or water injection volumes required to meet reservoir unit pressure constraints.
4) It should be possible to apply real-life constraints and objectives at any point in the network (not only at the boundaries).
5) The model should have a choice of fluid descriptions:
   - Black oil fluid modeling, using a variety of industry-standard correlations

- Fully compositional fluid modeling, using a variety of industry-standard equations of state (EOS). Should also allow black-oil-to-compositional mapping routines to allow black oil input into a compositional PVT system
- Compositional tracking should be possible throughout the network.

6) It should be able to perform production simulation as well as optimization. Simulation determines the pressures, temperatures, and fluid flow rates within the production system, while optimization determines the most economic production strategy, subject to engineering or economic constraints.
7) The program should model coupled pressure and temperature—that is, it should be capable of modeling temperature change associated with compression or expansion of fluids.
8) It should be a software model that can be used in all phases of field life, from planning through development and operations. This feature should enable petroleum, production, facility, and other engineers to share the same integrated model of the field and perform critical design and analysis activities, such as the following:
   - Conceptual design in new developments
   - Equipment sizing, evaluation, and selection
   - Daily production optimization, on- or offline
   - Problem and bottleneck detection/diagnosis
   - Production forecasting
   - Reservoir management
   - Data management
9) It should have the flexibility to model the behavior of a single well or pipeline, through to the most complex of fields with thousands of wells. This flexibility should come along with simplicity and ease of use.
10) The network modeling software should have the following features:
    - Scalability—The software should solve networks of thousands of wells in a robust and efficient manner.
    - Topological complexity—Any degree of complexity should be handled by the software, including loops and branches as well as distribution and injection networks.
    - Total system modeling—Injection and production networks should be solved and optimized simultaneously.
11) The network model should have the capability to link to a material balance model and run optimization for various time steps including events—wells coming on- / offline, additional equipment being installed, etc.

12) In order to obtain well performances for production forecasting, the software should be able to link with the well modeling package, to run it in a batch mode for generation of well performance and lift curves for simulation.
13) It should be possible to perform detailed modeling of separators, pumps, and compressors, should the optimization problems need it.
14) It should be possible to copy and paste parts of networks and to make bulk data entry or editing.
15) It should be easy to download or view and analyze results after completing a run.

2.3.3 Objects to Be Included in the Model

At the commencement of the modeling work, the process flow diagram (PFD) and the piping and instrument diagram (P&ID) of the field are made available to the modeling engineers. The modeling team may also visit the field to gain first-hand experience before finalizing the model appearance. The goal is not to include all the components seen in the P&ID in the network model. Noting that the size and level of complication of the model can also affect the optimization time, a decision should be made on what goes into the model and how detailed it should be. An acceptable level of detail that is required for the operation and optimization of the field should be included as a start. The model resolution can be improved any time later. Detailed modeling will require detailed input data for calibration.

For example, even problems with a produced water treatment system can restrict oil production. Any one of the constraints—capacity limitation of produced water treatment, reduced injectivity of the disposal wells, or problems with the water injection pump—can limit the oil production. So it becomes necessary to model the produced water treatment system and the disposal wells with all the details. The model should have all the equipment items to be able to input the real life constraints of the field. Only then, the model can generate solutions that can be used in the field. In general, all the processes that can potentially affect oil production should be modeled accurately.

While building the model, the modeling team will request the engineering details of the pipeline and various other components that would become a part of the network model. This will be jointly reviewed and agreed upon by the modeling consultants and the engineers from the operating company.

Engineering performance specifications would be used for modeling individual items such as pumps, compressors, and heat exchangers. It is necessary to include detailed models for these objects if it would add value for the optimization.

For example, to model a pump, two options are available:

- A *simple model* of a pump defines the horsepower it requires to deliver a certain increase in head for a certain flow rate of liquid. With a simple pump object, knowing the volume of fluid and the head developed by the pump, the optimizer would determine the horsepower required for the service, taking into account the pump efficiency. So if the horsepower required is available from the motor, the optimizer will return a solution with the horsepower used for the application.

- A *detailed model* of the pump, on the other hand, will require details such as type of pump (centrifugal or positive displacement), the design speed of the pump, and the pump performance curve. If the pump will operate at different speeds, a family of pump curves for the operating range of speeds will be available. The results from the detailed model are based on the pump curves, hence more realistic.

There are instances where detailed modeling becomes necessary. The level of detail would be decided based on specific optimization requirements.

For vessels like separators and heat exchangers, physical models may not be needed. In other words, for a three-phase separator, it may be sufficient to include an engineering object for the separator with a single inlet and oil, gas, and water outlets. The operational range of the vessel, based on its design capacity, can be imposed as a constraint in the model. Details such as length and diameter of the vessel and residence time are required only for process modeling. However, multi-phase separators with separation efficiency can be modeled as required.

Similar approach may be sufficient for eductors (used in gas recovery systems), extraction towers, de-salters, etc., if present in treatment facilities. The design flow capacity of these items may be input as a constraint in the model. Usually processing fixtures downstream of an HP production separator are not considered in detail for production optimization of the field.

2.3.4 System Technology

Information Technology (IT) keeps evolving. Some IT platforms are not supported by manufacturers beyond a certain time period. Therefore, the optimization software that is built on a specific IT platform should be able to survive when the technology improves. The operator should be able to evolve the models to the new technology platform, when required. For this reason, the software should be built on the current and most versatile technology base. This will ensure that it does not become obsolete within a few years. Usually, reputable software companies keep track of this issue and upgrade their products as and when necessary, to maintain a competitive edge in the market.

2.3.5 Flexible Service Provider

Flexibility with the service provider is an important criterion for a large optimization project. After reviewing various optimization technologies available in the industry, the operator would select the one that will be most suitable for the asset operations. If there are, say, three major service providers who seem to satisfy this criterion, the engineers from the operating company would have in-depth discussions with all the three parties, to understand the strength and weakness of the products before making the selection. It is not about the cost of the license alone; it is about the work process the operator is willing to adopt. Before the operator commits to investing their time, it is important to ensure that it is a reliable path to move on.

The operator usually seeks to understand the background of the service provider:

- How old the company is
- How long the company has been in this business
- How many permanent employees the company has, and the level of expertise of the staff

The company that has developed the optimization software would normally employ their own consultants to execute such consulting projects. It is better to sign up with the company directly, because if the software needs an improvement while the project is in progress, they will be able to develop it or fix a bug and deliver it on time.

The presence of a local service office or representative adds value. Should there be a problem or delay, it can be easily communicated face to face. If the service company is on the other side of the globe, with a significant time difference, it may be difficult to get immediate help. This may not be an issue, with the communication technology that is available today.

It is not necessary that the service provider should be the largest service company in the oil and gas industry. It is the technology and the industry presence that should be at acceptable levels. The service provider should have a reasonable market share and a competent track record. If the company is nascent, there is a big risk in hiring their services.

The service company personnel should demonstrate a willingness to succeed, as every optimization project is different. The flexibility demonstrated by the service provider will be a key factor for the operator in selecting and bringing them on board to work as technical partners.

Another factor the operator should consider is cost. A system should not be selected because it is cheap. A cheap optimization system may satisfy the near-term requirement, but may not be able to handle the future vision of the asset. Selecting a system that does not have a future scope for expansion would be the wrong decision for the long term. As an example, if the first aim is to optimize wells, the operator would build models for hundreds of their wells. But if the well models cannot be progressed to a system optimization or real-time optimization in the future, it will become necessary to rebuild all the well models again with a new software. This is very inefficient.

The system selected should have smooth connectivity with old and new databases of the operating company—be it the legacy database systems or recently set up real-time data acquisition systems. Data transfer should happen seamlessly. This should be considered up front so that this does not become an obstacle in the future. Usually, operating companies are reluctant to change their legacy data systems. A workaround should be invented.

2.3.6 Third Party Feedback

While every service provider may claim that their product offers the best-in-class solutions, the operating company should work objectively to ascertain if the product selected is the right one for the asset.

It would be better to verify that the proposed technology is being used for modeling and optimization of similar-sized networks elsewhere. The operator could ask for references. If the service provider confirms that their software has been used for the optimization of similar, complex multi-phase networks, the operator could ask for the contact details of a senior engineer of the operating company where a similar model has been deployed. It is usually possible to communicate with the operating company engineer to get the full details of the project within the confidentiality limits of the company.

It is preferable (although not always possible) to get feedback from the staff of two or three operating companies who are using the product. The operator should seek answers from the user companies to questions such as the following:

a) What are the strengths and weaknesses of the product, based on their own experience?

b) Did the product and the process deliver the minimum benefits they had targeted for?

c) Is the operating company satisfied that they selected this product? Was it worth the effort?

d) What did they do to really achieve success and acceptance in their organization?

This will verify the claims made by the service provider and will also assist the operator in setting realistic expectations for their project.

2.4 System Optimization Implementation

2.4.1 Implementation Steps

The typical implementation of system optimization in an asset consists of the following twelve steps:

1) **Decision to develop an optimization system**

Usually, the operating company decides to gain the benefits of system optimization and floats an enquiry. A brief description of the process flow of the asset and the scope of work is included as a part of the tender enquiry.

2) **Vendor selection**

There are formal meetings as various service providers want to understand the current status of the asset—especially the level of instrumentation and optimization workflow already existing in the asset and details of the optimization objectives. Some operators invite the service providers to come and make presentations so that the technical team of the operating company gets a clear understanding of the offering from the service company. The tender evaluation is completed and the offer awarded to the successful bidder, who has been selected based on the merits of their technical and commercial proposals. Usually, a work plan, deliverables, and project schedule are committed in the tender proposal. A contract is signed with the company that will provide the modeling services.

3) **Kick-off meeting**

The kick-off meeting plays a vital role and is crucial for the project. Many team members get a realistic understanding of the project and their role only during the kick-off meeting. They realize the quantity of work that needs to be done on either side before they can expect some results. All the stakeholders should attend this meeting because the consultants will really start the work based on the outcome of the kick-off meeting. It is very important to include a prominent member from field operations on the team. The consultants share their intended work plan and seek assistance from the operating company to help them build a quality model and establish a best-in-class optimization process. If someone has some concerns about the data quality, work plan, suitability of the software, or the details of the deliverables, it should be brought up in this meeting and not later. The single point of contact (SPOC) from the operating company and the service company will be assigned.

Another decision that is agreed upon during the kick-off meeting is where the project work will be executed. The consultants may prefer to work from their office, subject to the approval of the client. In some cases, the operating company will insist that the work should be done within their premises, due to policies of data security and confidentiality; in this case, they provide work desks for the consultants in the office during the project period. In either case, it should be possible for the staff of both client and the consultancy firm to freely interact with each other to share information and review work progress informally, as the project progresses.

Issues such as data security, confidentiality, and service desk requirements for the consultants are discussed. The minutes of the kick-off meeting are documented, circulated, and agreed. This becomes a part of the contract.

4) **Data collection and QC**

   The consultants provide a detailed list of data required for the modeling work. This data should be either pulled from an existing database, collected from field operations, scanned from hard copies, or assembled new, as requested, in Excel or similar format and provided to the consultant. The main communication will be between the SPOCs of each side. This way, the operating company has a record of the data shared with the consultants. The consultant also has a written record of what he or she has asked for and what is still pending. There is communication back and forth about the quality and consistency of the data. The parties finally agree on what data is to be used and which items to be assumed. This is also documented.

5) **Modeling work and progress reports**

   As the modeling progresses, the consultant SPOC shares the work progress with the operating company SPOC. This is actually for information as well as validation. The consultants may also request one or two field visits to see the field operations and clarify their doubts, if any. A field visit gives the consultants a good level of confidence on which modeling choices to make. This could be followed up through emails, phone calls, and video calls. Regular progress reports are submitted by the consultant (weekly or fortnightly). The reports include the activities carried out, the percentage progress (what has been done till now), and if any important data or clarification is awaited from the client.

6) **General training session**

   This is conducted by specialists from the consulting company. The objective is to expose the optimization software to a large group of people from the operating company. The consultants conduct detailed training about the optimization process and capabilities and features of the optimization software. Temporary software licenses are given to the participants, and they are encouraged to "play" with some models to become familiar with the model building activities. This session also helps to spread the awareness and strengthen the expectations of the optimization effort within the operating company.

7) **Model history match**

   Once the model is about 90% complete, the consultant asks the operating company SPOC for one full set of process data on a quiet day (a day that had no process upsets, with almost all wells flowing and all compressors running). The consultants complete the modeling work, perform history match using the field data, debug the model, analyze results and understand the model, and prepare reports.

8) **Preparation for optimization**

   The consultants meet with the operating company and share the results of the key findings of the history match. They also seek input from the customer regarding the optimization scenarios. The operating company SPOC discusses with his teams and spells out about four or more different scenarios that should really add value to them. It would be better to select at least one or two scenarios that they had been seeking a solution for in the past but could not yet get, as they had no network model before.

9) **Optimization runs**

   Consultants set up and run the optimization scenarios. They debug the model and ensure good and stable convergence, analyze the results for each case, and frame the conclusions from the model results for each optimization scenario. They compare the results and understand the real constraints in the field and unearth any hidden potential that can be realized without major capital investment.

10) **Results presentation**

    Consultants present the results of the history match and optimization scenarios to the main team in the operating company in this important meeting. All stakeholders attend this meeting. Consultants take feedback from the attendees and submit a draft plan for continuous optimization in the future, considering the organizational setup of the operator. They may also suggest possible changes in the organization structure, for the operator to reap benefits from continuous system optimization.

11) **Advanced optimization workshop**

    The workshop is conducted for a smaller group of "core" engineers and operations staff who would use the model on a regular basis. The actual model to be delivered is used in the workshop. They also get feedback from workshop attendees. The model is refined, as needed. A draft project report will be submitted for review.

12) **Project sign-off**

    The consultants finalize the models and complete their analysis of the results, update the reports with final conclusions and recommendations, get a peer review done, deliver the history match and optimization models with reports, and sign off on the project.

Beyond this point, any follow-up visit and work would be done by the consultants as agreed in the contract or based on a revised work order from the operator.

If the asset has many fields, the operating company may choose to do a pilot optimization project first for one of their fields. Depending on the outcome of the pilot project, they may decide to extend the project to the other fields. In either case, the implementation process will cover all the steps mentioned above.

The interaction between the consultants from the service provider and the staff of the operating company is the most important aspect of this work, which increases the value of the project for many years to come. The execution plan highlighting this interaction between the operator and service provider is shown in the chart below (**Fig. 2-14**).

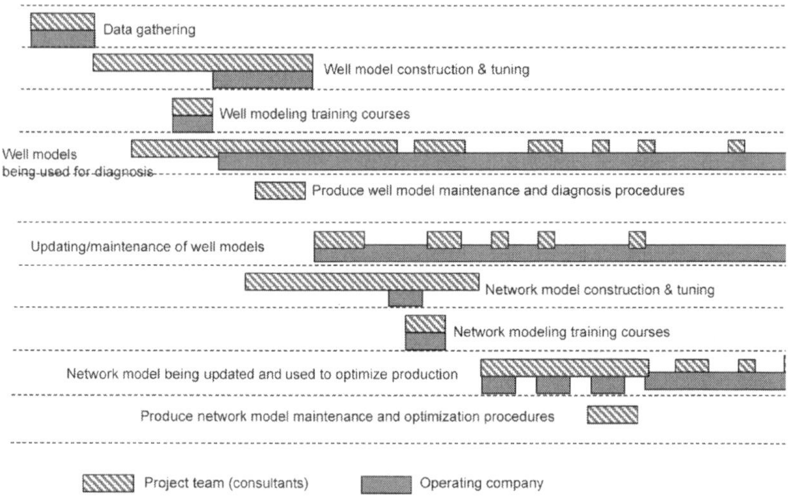

Fig. 2–14. Execution plan—offline modeling

After establishing offline modeling and optimization, it takes some time for the operating company to imbibe the workflows. There could be minor issues with the model that the optimization engineers should get a handle on, with the help of the modeling consultants. The confidence in the model will grow as the optimization team uses it for several in-house optimization evaluations.

The offline system optimization has been implemented successfully in assets in various parts of the globe and is well documented. [(8), (9), (10), (11), (12), (13), (14)]

2.4.2 Data Required for Network Modeling

A general list of data required for network modeling can be found in **Appendix A2.** [(15) (16)]

2.4.3 Tangible Benefits

The system optimization is a heavily involved work from the operating and consulting companies. Most importantly, it needs extra time from the engineers of the operating company. These engineers are not free—they normally do their routine work and handle the system optimization model responsibilities as additional work. Therefore the prize or the benefit should be attractive, tangible, and measurable for the operating company to engage its workforce in such an activity. If it is not measurable, people would not believe in it.

In typical gas lift system optimization projects across the globe, the following benefits have been realized:

- An increase of 2–5% oil production
- A saving of 5–15% of lift gas, plus
- Many other benefits

In many of the optimization projects undertaken by the author, in assets with moderate water cut, a 2–5% increase in oil production rate has been practically achieved. In old assets with high water cuts, an increase in oil production of 1–2% has been realized.

The cost levied by the consulting company for the project consists of two components: the consultant fees and the cost of software licenses. Both are one-time payments. The cost of licenses will increase with the number of concurrent users. The number of licenses for concurrent use of the software depends on the number of users and number of fields, and also on the company's business requirement.

From the second year onwards, there would be a maintenance fee (a small percentage of the cost of the license fee), payable annually, for the software company to provide regular software upgrades.

For smaller fields (say, with less than 25 wells in the field), a less expensive, *lighter version* of the license can be used. A lighter version will be cheaper, but may limit the number of wells or pieces of equipment in the model. This would keep the project costs lower.

The value of the increase in production is significant. If the field production is 50,000 BOPD and if the project increases production by 2%, at a price of $30/BBL, the value added is (1000*30*30=) is $900,000/month. This gives an idea of the fast payback for the project.

There are other intangible benefits—using all the people skills to contribute their best while working in a team environment to achieve the best human potential, avoiding redundant work, reducing the well or equipment downtime, making operation transparent, improving accountability of people, etc. These have high value, especially for the long term, although not easily quantifiable.

2.4.4 Contributing Factors for Success

Apart from factors such as optimizer technology, data quality, and skill sets of the people involved, in many projects, the main factor that makes a big difference and ushers in success is the commitment and passion of people who are involved in the project, both from the service and the operating companies. The creativity, flexible thinking, and willingness to change things demonstrated by the operating company makes a big difference. It is the role of management to identify such people who have a passion for optimization and involve them from the very start of the project. Without the active involvement and contribution of the operating company personnel, the consultants or the software alone cannot unlock the full potential of the field.

2.4.5 Model Maintenance Protocol

In any organization a protocol should be established for maintaining the network optimization models without getting them corrupted or deleted. The models are usually kept on servers, and these servers are backed up on a daily basis. The rules would be set within the company that, even though many users may have access to the software license to open the model, the users of the model have different privileges for security reasons, as shown in **Table 2-3.**

Table 2–3. Model user privileges

User	Privileges
Network Model Administrator (1 with backup).	Has privileges to download and upload models into the central repository (save, change, or delete and replace the master model from the network).
Power users (2-4 users).	Can download the model from the network to their workstations and make changes to the model and converge it, and prepare for re-insertion onto the server. But will request NMA to replace the existing model with the revised model.
Normal users (up to 10 users).	Can download models from the network and view the results from the converged models in their work stations. Out of interest, the normal users can reset the results and re-run the models with different parameters, should they wish, but usually this would not be required. They can get this done officially through power users on their teams.
General access to view in text format	This enables many members of the asset team to have access to view the data through the intranet; they can view the data as well as model results at key locations in text format. They do not need to access the model for this purpose.

The license may reside on the server, but the program is usually installed on the individual desktop connected to the network of the client. This will aid fast loading of the program and quick opening of the model, once the license is checked out. It is advised that the user should download the model from the network server and work from his own workstation PC so that the communication path between his PC and server is not jammed while working on the model.

The number of licenses can be decided based on the usage of the software within the organization. For this purpose, the network administrator can monitor the license usage for a few months, and a decision can be made whether additional licenses are required.

2.4.6 Scenario after Project Handover

It is common thinking that optimization is a one-time effort. This is not true. As the field changes continuously, the optimization too should be continuous. The frequency of optimization would depend on the business requirements and the size and complexity of the field. A large field with several hundred wells takes more time to stabilize after a change. Significant time will be required to collect a full set of field data if the data collection is done manually.

Some companies initially incline towards employing a Design and Evaluation Services for Clients (DESC) engineer from the consulting company as a part of their team; they use his or her special skills to take optimization forward. As observed in many assets, this approach may work for a short period (say, up to one year) but cannot continue forever.

The best option for the operator is to become self-reliant and take ownership of the optimization process. If there are some convergence issues with the model or if the model needs some upgrade or a major revision (e.g., addition of a new field to the model), the operator can utilize the service of a consultant engineer for a short period of time for the specific task. There would be a provision for this type of call-out service in the original contract. In many fields, over a period of time, the senior optimization engineers of the operating company become experts in the software and the model, because of their powerful hands-on experience with the model.

2.4.7 Routine Model Update

The biggest challenge for any field is to be able to continue the journey of system optimization over time. The model should be updated to represent the current conditions of the field before performing any optimization run.

As operations continue, the flowing characteristics of wells change with time. Some wells get shut in, and new wells come online. Produced volumes of fluids and ratios such as water cut and GLR change. Different pieces of equipment are pressed into service. Production lineup might change. Pipeline conditions will become different due to corrosion or solid / wax / scale build-up. Key operating parameters of the field might change. Keeping these in mind, the model should be updated and history-matched, if needed, before performing fresh optimization studies.

The most time-consuming aspect of updating a network model is re-calibration of well models. This involves a lot of work in assets with hundreds of wells.

The engineer uses the most recent well test data and performs nodal analysis to retune the well model. It may require adjustments to be made to the inflow or outflow parameters, or both, to get a good match for the measured production rate and flowing bottom-hole pressure.

Fig. 2-15 shows the different adjustments that may be made to: (A) inflow, (B) outflow, and (C) both inflow and outflow, in order to obtain a model match with the latest production rate (the dashed lines show the adjusted parameters).

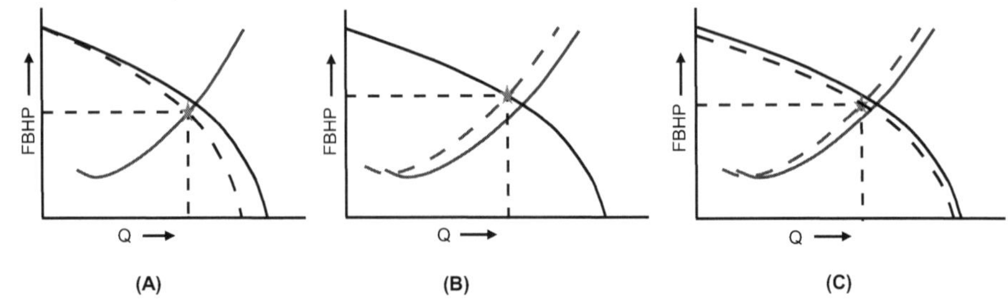

Fig. 2–15. Recalibration of a well model with a new well test

As the engineer-in-charge regularly monitors the well performance and is familiar with the performance trends of the wells in the area, making the adjustments to the model is not difficult. If a reasonable match is not obtained with minor adjustments, the engineer initiates additional investigation that might include downhole pressure and temperature surveys.

Upon successful recalibration of the well model, a new well performance curve will be generated. This will replace the existing performance curve for the well in the network model.

In the past, the steps such as well test validation, well model calibration, performance curve generation, and updating the network model with the new performance curve were done manually. Today, in many assets, these steps are performed and managed by RTO systems.

2.4.8 Integrated Asset Optimization Model

The network model referred to so far does not have a direct link to the reservoir, as the inflow part of the well models represents the reservoir component at any time. However, while creating an asset development plan, the following details are considered:

- Oil production targets vs. time
- An approximate drilling plan—how many wells to be drilled over the years
- Water / gas injection plan—to maintain reservoir pressure and improve sweep efficiency
- Commencement of artificial lift (gas lift, in this case)
- Timeline for expanding gas lift operations (additional compressor, separator, etc.)

- Limits and economics of primary, secondary, and tertiary recovery plans
- Field abandonment

It is indeed necessary to understand the conditions of the reservoir, wells, and the surface network together *at any time* to decide what additional facility will be required and when it will be required.

This holistic approach that links all three components—*reservoir, wells,* and the *surface network*—is called an *integrated asset optimization model* (IAOM) or *integrated production model* (IPM). This can be used to forecast field production and to identify the timing of the addition of wells and facilities expansion in the field. The integrated modeling approach is desirable, as it combines the surface and sub-surface assets in its totality.

For this reason, it is necessary to link the surface network model to the reservoir simulation model. The reservoir model can be an existing reservoir simulator or a tank model based on material balance.

It should, however, be understood that the network model follows a fast loop optimization, whereas the reservoir follows a slow loop optimization. Thus, if the network model is run connected to the reservoir component, it will take more time to return a solution.

Stand-alone production optimization runs of the network model for day-to-day optimization will require short duration only (say, 15 minutes or less, depending on the configuration and size of the field, number of constraints handled in the model and the robustness of the optimizer), but when connected to the reservoir model, it will take several hours.

For a day-to-day optimization, a link to the reservoir component does not add any value. The calibrated well flow models prescribe the inflow capability of the reservoir. The system optimization linked to a detailed reservoir simulation adds value only for long-term production forecasting.

Another simple way of approaching this is to run time steps of the reservoir model and link the output files to the surface network model. Details of the time-based reservoir work program, such as how many new producers and injectors will be drilled and added to the field and how many wells will get shut in, etc., can be included in the stepwise optimization. The time step can be once a year or once in six months. The network model would show the limitations it would face in those future conditions. This would indicate optimal timing for procurement and installation of additional equipment and facilities expansion in the field.

2.4.9 Variety of Uses of a System Model

A system optimization model serves a wide variety of purposes, from conceptual development to abandonment of the oilfield. The dominating benefit of a network model is production optimization during different phases of the operating life of the field. The many ways in which a network model assists operations are listed below:

1) Before commencing production, the network model can be used to predict production from the field, considering the practical limitations of sub-surface and surface components. This is useful for proper sizing of the equipment and pipelines.
2) Once field operations begin, optimization evaluations suggest the best possible solutions to maximize the profit from the field, while running operations with existing process constraints (e.g., water handling limitations, pipeline pressure limitations, etc.) and achieving conformance to contractual requirements (e.g., limits on impurities or inorganics such as CO_2, N_2, and H_2S in export gas). The price of oil and gas and the cost of operations (cost of water treatment, field maintenance, and power) can be accurately factored into the model while performing system optimization. The final outcome is the mode of operation with reduced Opex.
3) A network model can assist in the proper gas lift distribution in a gas lifted field. This is extremely useful during normal operations, as well as during compressor maintenance. As the gas lift and production networks are optimized together, superior results are obtained for maximizing production through effective utilization of lift gas.
4) In complex operating environments with several operational possibilities (gas export/import, supply to gas lift, supply gas for reservoir injection, compressor shut-down, etc.), a system optimization model gives the best techno-commercial solution on a daily basis.
5) The sub-optimal performance of any component of a network, either surface or downhole, can be identified using the network model. The model can also predict the impact of improving the component performance. This helps operations to get an estimate of the expected benefits even before initiating the remedial work.
6) The simulation of the field network would identify pipelines with serious problems such as plugging, solid dropout, and water loading that may exist in any part of the network. The most congested pipeline in the system can be identified by the network model, thus helping to perform de-bottlenecking in the field.
7) If the field operation efficiency is affected by a piece of equipment by design or performance, the model can suggest the specification of the optimum equipment that would be ideal for the current and future operating conditions.
8) A network model is extremely useful in evaluating development project options. If several project options are possible, these options can be

evaluated using a network model. The model results will indicate the best project option that leads to optimal capex utilization.

9) The network model can be used to allocate field production back to the wells. The downtime of wells can also be factored in the back allocation.

10) The network model can be used for production forecasting. Either the well model parameters should be changed for various time steps, or the network model should be connected to a reservoir model to forecast production. The forecast is more accurate as both the reservoir behavior and surface pipeline / equipment constraints are considered together.

The diverse uses of a network optimization model is captured in **Fig. 2-16.** [3]

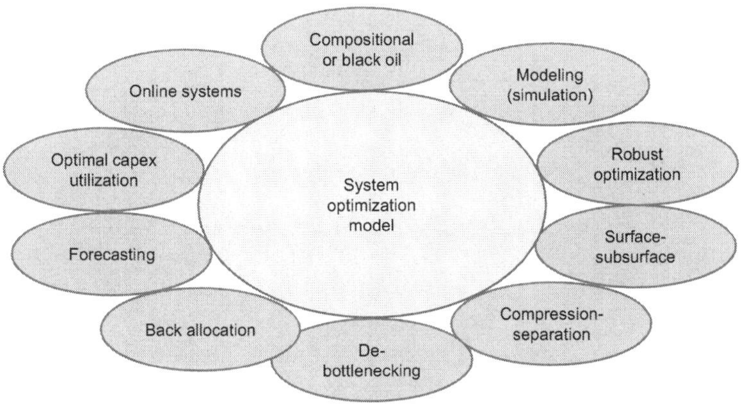

Fig. 2–16. Diversity of a system model

2.4.10 Total System Optimization—Benefits and Concerns

System optimization is a result of multi-disciplinary interaction. A senior petroleum engineer or optimization engineer usually leads the effort. Several disciplines such as production engineering, production operations, reservoir engineering, rotating equipment engineering, pipeline engineering, and systems engineering work together in the optimization process. A true field optimization cannot be achieved without active participation of various teams.

The advantages or benefits of the system optimization model are summarized below:

1. The most valuable benefit of system optimization is that it will make the people work together. After embracing the system optimization approach, even in what used to be *silo* organizations, significant improvement is seen in the work culture. Members come forward to share data and work together to achieve the common objectives. They enjoy this way of working.

2. The beauty of the system optimization approach is that once a calibrated network model is available, several scenarios can be investigated in a short period of time. Any suggested mode of operation can be investigated. The scenarios can be compared, results reviewed, and shared with other teams as required. This helps various stakeholders review the results of several options quickly and arrive at a consensus. The best way forward becomes apparent to all concerned.
3. These models are extremely valuable in making decisions about new projects that require capital investments. The system model can be used to prove or disprove the viability of several big projects "floating in the corridor." Typical examples are projects that require significant capital investment—an additional compressor or separator or a new subsea line.

   As an example, if a marginal field is discovered adjacent to an existing operations, an important question that should be answered is: should the production from the new field be tied to the existing field facilities, or should facilities be built for the new field? The new field can be brought on-stream quickly if tied to the existing process facilities. It might have some cost savings now, but it will put constraints on both the fields forever. Building new facilities will be good for future expansion, but will cost more money and will delay the commencement of production. The production impact of these two options can be readily studied with the help of a system optimization model. With these results, the best business case can be justified and implemented.
4. While sanctioning projects, partners and shareholders always look for project economics and justification, showing the cost versus benefit numbers generated on a sound technical basis. The system optimization model offers a technical basis for a project proposal.
5. The network optimization approach makes it possible to use flexibilities associated with existing wells, pipelines, and process equipment to achieve better business results. Sometimes optimization runs unearth hidden potential, hitherto disregarded and not considered to be value-adding.
6. If there are several possible technical solutions to a field problem, the merits of each solution can be investigated quickly using a network optimization model. This can be repeated for all options and the results compared.
7. If there is an opportunity to use a new technology, the existing model can be modified to incorporate the new process or device, and its impact on the field performance can be evaluated using the model. This is an easy way to justify new technology, as some people might be apprehensive about the use of new technology.

8. The field network model connected to a reservoir simulation model would aid accurate production forecasting of the asset.

The risks or concerns associated with the use of a system optimization model are summarized below:

1. In the absence of a fully committed multi-disciplinary team, the system optimization model will not yield the anticipated benefits.
2. The optimization software chosen should be the right one for the application on hand. The optimization technology should be proven and scalable (i.e., should be able to add any number of wells to the network). The technology should be selected after rigorous investigation by the operating company.
3. Good-quality data is a requirement for successful system optimization. Getting data in some fields could be a challenge, as the asset may not be fully instrumented. The operating company must make provisions for the data collection as required. Live digital data is the ultimate target. The investment required for fetching live data should be factored in.
4. Modeling requires highly skilled engineers who have overall knowledge about all operations taking place in the field. The operating company should involve people with this skill set in the modeling project from the very beginning and ensure that sufficient in-house skills are developed before the project is handed over. If this is not done, the project is likely to fail.
5. Even though a big portion of the initial work is done by consulting company engineers, participation of the operating company is critical during the project execution phase, as well as after the project handover. The human power required from the operating company should not be underestimated. Training and knowledge transfer to the operating company staff are key factors.
6. The biggest challenge is to maintain and update the model as the field undergoes changes. This requires a considerable team effort. If the model is not current, the results will not be relevant to the present operations. This will make the model obsolete and unusable.

3

Modeling of Network Components

All types of wells, pipelines, and equipment available in the field are included in a comprehensive network model. The level of detail incorporated for every component is decided jointly by the modeling specialists and the operator representatives. This will ensure that the model can simulate the constraints encountered in the field and be able to perform a variety of investigations accurately.

The following topics are discussed in this chapter:

- Model appearance, naming, and color-coding conventions
- Fluid modeling for a network model
- Incorporating well performance in a network model
- Modeling of components such as pipelines, fittings, pumps, compressors, and separators
- Detailed modeling of centrifugal compressors

3.1 Model Appearance and Naming of Components

Even though modeling requires a thorough understanding of all the production processes in the field and the working principles of various types of equipment such as pumps and compressors, the final product—the model itself—should be easy to use. This is important because all the people who will view the results of the model for field implementation or project discussion may not have the same level of expertise as the person who built the model. If the model is constructed in an easy-to-use manner, the usage of the model by operations and other teams will increase. This is the desired result.

The appearance of the model on the user screen is as important as its technical strength. Even if the model performs well and gives reliable solutions, if it is not constructed in an easy-to-use manner, the average user may lose interest while trying to use the model. The user should be able to follow the process within the model easily and extract the results quickly. If someone wants to see the results of a compressor on a platform for a particular scenario, the user should be able to navigate to that page quickly in a logical manner and be able to access the unit.

The model may have thousands of pipelines and lots of process equipment, yet it should not look congested or cluttered. It should be organized well so that anyone can work with the model easily.

This can be accomplished by following some simple tips, described below.

3.1.1 Using Colors to Differentiate Lines

Different colors may be used for different process fluid lines in the model. Some operators have their own standards for this within their organization already, so if the same color scheme is followed in the network model, it becomes easy for the users. For example:

- Red—High-pressure gas lines
- Light brown—Low-pressure gas lines
- Blue—Water lines
- Green—Oil lines
- Black (default)—Multiphase fluid lines

3.1.2 Layered Display of the Model

The most convenient way for displaying the full model on a screen is by layering, one underneath the other, using sub-sheets. If the model is opened, the aerial or top view of the field should appear on the first screen. This sheet should allow the user to drill down to any location or component the user wants to see.

A simple example of an onshore field with two main production clusters and one central production facility (called CDS—central degassing station) is shown in **Fig. 3-1**.

Each cluster has a production gathering manifold and test separator. The produced fluids from the clusters flow to the CDS. There may be wells around the CDS that may flow directly into the manifold in the CDS itself. The major items such as production separators, main oil pumps, gas compressors, water treatment, water injection pumps, and other utilities are located in the CDS.

If the user is interested in knowing about the condition of a well in Cluster-1, upon opening the model, the user should:

- Click on Cluster-1; the screen will display a high level overview of the cluster
- Go to the wells area of the cluster and click on the particular well (the model will display the wellhead object connected to the network and show if the well is open or shut-in)
- Click on the wellhead object to view its performance curve (if the model already has a converged solution, the result will also be marked on the performance curve)

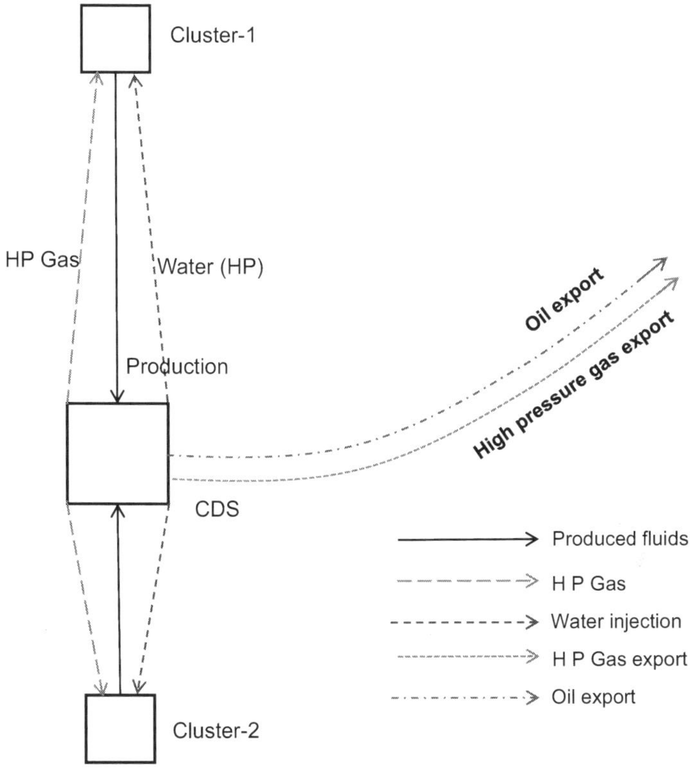

Fig. 3–1. Model view of onshore field layout

Alternatively, the user will be able to search and find the well using a "find" option similar to the MS Office package, as long as the well number is known. For this reason, the modeling consultant should use a uniform and consistent way of naming the wells; e.g., well number 15 in Cluster AP is named *AP15*. The same naming scheme should be followed universally throughout the model.

3.1.3 Naming of Pipelines and Equipment

The pipelines and equipment in the model are arranged to appear similar to how they are physically located in the field. It is not drawn to scale. Most of the users have been there previously or have seen the location maps of the production clusters or offshore platforms many times, so making the arrangement in the model appear similar to the physical layout in the field is advantageous.

An arrangement of a small offshore field with five satellite platforms and a central facilities platform is shown in **Fig. 3-2**.

As can be seen, platforms A and D have gas lifted and water injection wells, whereas platforms B, C, and E have only naturally flowing wells.

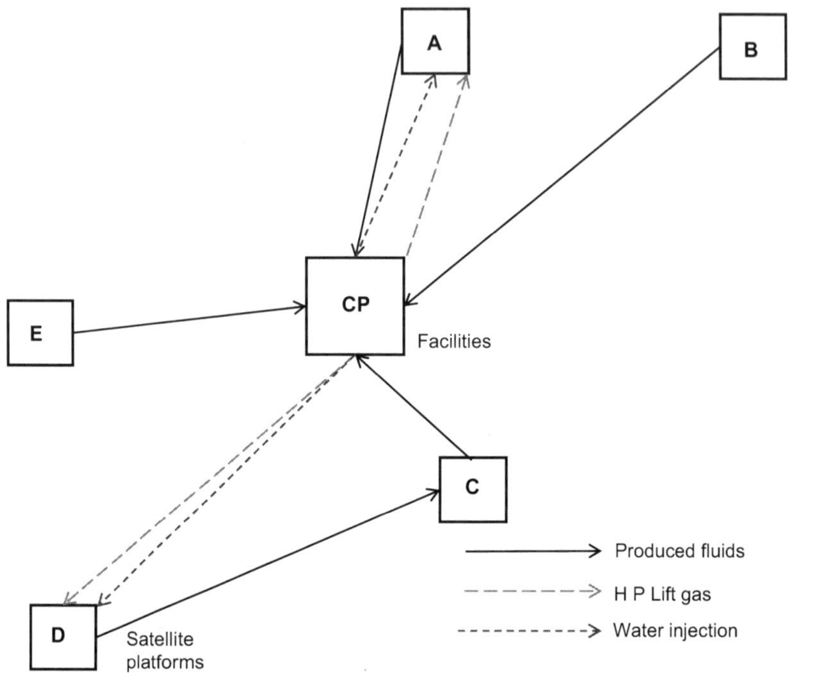

Fig. 3–2. Model view of offshore field layout

When naming the pipelines, a simple, easy-to-follow naming convention is followed.

- The line named *CP-A-12-GL* represents the pipeline from Central Processing to Platform-A on the gas lift service, nominal size 12".
- The line named *D-C-18-Prod* represents the pipeline from Platform-D to C, and *C-CP-24-Prod* is the name of the pipeline from Platform-C to Central Processing, carrying the produced fluids.

3.2 PVT Modeling

3.2.1 Fluid Modeling for Wells

The term *fluid model* refers to a single fluid, which can be defined in one of two ways:

- *Black oil data*: The fluid is defined in terms of stock tank oil and gas gravities and a separator gas-oil ratio (GOR or CGR).
- *Compositional data:* The composition of the fluid is defined using components (e.g., C_1, C_2, C_3, H_2S, etc.) and their properties (e.g., molecular weight).

The types of fluids to expect from a reservoir (black oil, volatile oil, gas condensate, or dry gas) are determined through downhole fluid sampling or offset well data. Detailed individual fluid modeling is required for well models.

The black oil approach to fluid modeling is simple and widely used in the petroleum industry. When using black oil data for modeling, fluid models can be divided into three different types:

- *Black oil and volatile oil:* Defined by entering oil and gas gravities and GOR
- *Condensate and wet gas:* Defined by entering condensate and gas gravities and CGR
- *Dry gas:* Defined by entering gas gravity only (fluid is treated as single-phase gas at all pressures and temperatures)

The key PVT properties to be modeled for each formation fluid are listed below:

1) Specific gravity or density
2) Bubble point pressure (indicating when the first gas bubble comes out)—this is applicable for oil and volatile systems.
3) Dew point pressure (the point at which the first liquid droplet forms)—this is applicable for condensate and dry gas systems.
4) Solution gas-oil ratio (R_s)—describes the quantity of gas dissolved in an oil at reservoir conditions
5) Formation volume factor (B)—shows how much shrinkage happens between reservoir and surface
    - B_o—volume of reservoir oil required to produce one barrel of oil in the stock tank
    - B_w—volume of reservoir water required to produce one barrel of water in the stock tank
    - B_g—volume of reservoir gas required to produce one scf of gas in the stock tank
6) Viscosity (μ)—a measure of how much resistance it offers to flow
7) Surface tension (σ)—helpful in flow regime and holdup determination, even though the effect is small.
8) Compressibility (z)—deviation from ideal gas behavior

Inorganic components such as N_2, H_2S and CO_2 are entered as mole percentages, as this is how they are generally expressed in PVT reports. Water content may be expressed as a water cut, WGR, or mole percentage, depending upon the context in which the data is to be used.

In the black oil description, the fluid is modeled using *correlations* that predict the behavior of R_s, P_b, B_o, μ_o, μ_g, and Z_g at different pressure and temperature conditions. The measured data is input and tuned using applicable correlations in the software. Condensates and volatile oil are more accurately modeled in the modified black oil description using P_d (dew point pressure) and R_v (vaporized oil rate) behavior.

A black oil PVT approach is usually sufficient for black oil and dry gas, hence tuned correlations are used for modeling these fluids. However, for gas condensate and volatile oil systems, compositional modeling using equations of state (EOSs) is preferred.

3.2.2 Fluid Modeling for Network

For the network modeling, a *fluid source group* (FSG) is created and used. This is a family of fluid models tuned together. An FSG replicates the production situation in the field, as different formation fluids flow and mix in a manifold and other parts of the network before arriving at the separator. Each FSG can be of either a black oil or a compositional description.

Usually, software providers have one common fluid modeling software that would be fully compatible with the well and network modeling and reservoir forecasting packages.

If five wells producing fluids from five different formations flow into a network, these five fluids should be merged in a single FSG. These individual fluids would have been already modeled and tuned in the PVT modeling software during well modeling. If there are other fluids that are handled in the network model, e.g., lift gas or injection water, these fluid models also should be added into the fluid source group. Once all the fluid models are added, a group tuning is performed. The resultant fluid group is saved with a name (e.g., FSG01) and assigned to the network model.

Even though the FSG is assigned to the network model, each fluid model from the group will be assigned to the wellhead or source object handling that fluid. This will allow the network model to process the mixing of fluids at manifolds as it happens in the field.

Only one FSG can be assigned to a network model at a time. This means that for running a different optimization scenario with the same network model, the user can de-assign the previously assigned FSG and assign a new FSG, if needed (e.g., FSG02).

This modeling methodology is used in the system optimization suite provided by Weatherford. Other products also follow a very similar approach.

3.2.3 Black Oil Fluid Source Group

In a black oil source group, the fluids—dry gas, gas condensate, or oil—will be defined using experimental data, and correlations will be selected for the group for tuning. Fluid properties from PVT reports will be something similar to the display in **Fig. 3-3**.

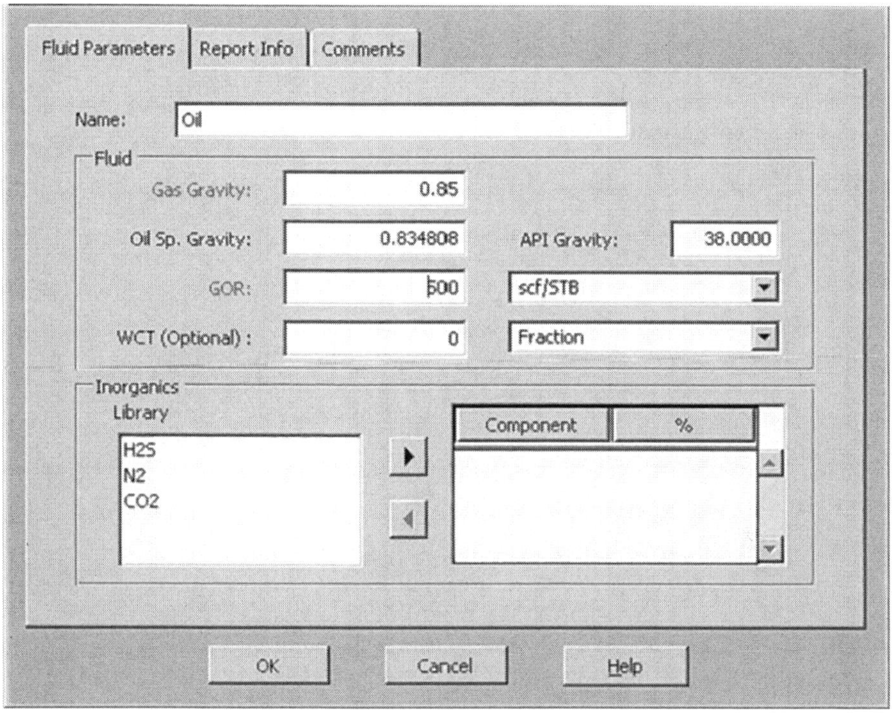

Fig. 3–3. Oil fluid model properties

The following are some of the commonly used PVT correlations available for tuning black oil fluid models (**Table 3-1**). The correlation that is most suitable for the oil sample should be chosen for every fluid:

Table 3–1. Correlations for black oil fluid models

Bubble point and Solution GOR (P_b, R_s)	Oil formation volume factor (B_o)	Oil viscosity tuning factors (μ_o)	Gas viscosity and Z-factor (μ_g and Z_g)
Standing	Standing	Beggs	*Gas viscosity*
Lasater	Lasater	Beal and Chew	Lee
Vasquez – Beggs	Vasquez – Beggs	ASTM (only for dead oil)	Carr
Glaso	Glaso		*Gas Z-factor*
Petrosky-Farshad	Petrosky-Farshad		Drunchuk et al.
Macary	Macary		
Marruffo (P_d and R_v)			

Note that different correlations may be used for tuning each fluid model, but only one correlation can be chosen for each group while tuning the FSG. Once all the black oil fluid data is entered and correlations selected, tuning is performed as shown in **Fig. 3-4**. The tuning factors obtained apply to the fluid source group as a whole, and not now to individual fluid models.

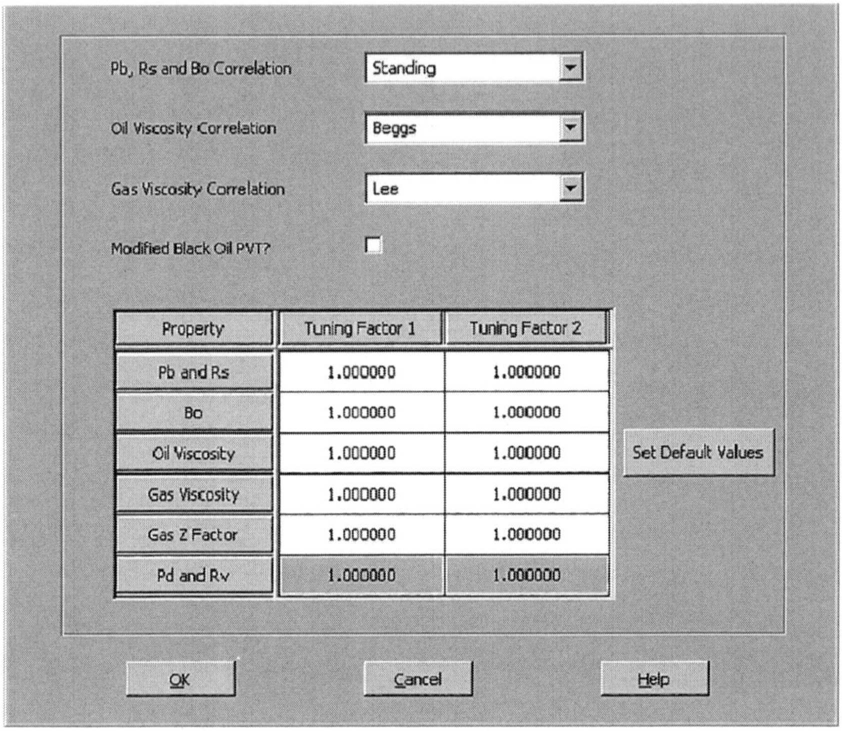

Fig. 3–4. Black oil correlation selection and tuning for FSG

For the example in **Fig. 3-4** the following correlations have been selected for the FSG:

P_b, R_s and B_o: Standing

μ_o: Beggs

μ_g: Lee

Suppose wells from two formations are being modeled in a network model, and 80% of the fluid comes from reservoir A and the balance comes from reservoir B. When tuning the FSG, we would preferentially use the correlations for R_s, P_b, B_o, μ_o, μ_g, and Z_g that are more suitable for the reservoir A fluid, as this fluid will dominate in the surface network.

Additional wells that produce fluids from another reservoir C may be added to the network in the future. When these wells come online, the fluid split in the network will be different. With three fluid models, the correlation selection for the tuning of the FSG may change. For this purpose, several FSGs may be created and kept on standby, even though only one FSG will be used for one snapshot. The key is to assign the most appropriate FSG for any simulation or optimization run.

The black oil FSG is constructed and used in a network model, as shown in **Fig. 3-5**.

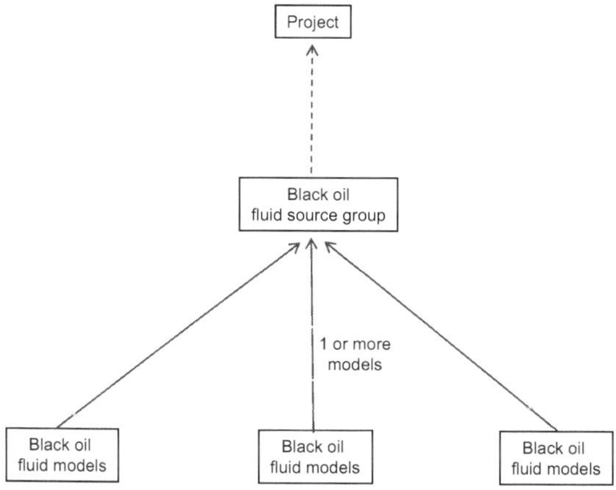

Fig. 3–5. Black oil FSG construction

3.2.4 Compositional Fluid Source Group

Compositional fluid modeling is a much more complex, and more accurate, method of describing fluid behavior.

In building a *compositional fluid source group,* the following are used:

- A single equation of state (EOS)
- A single, common plus-fraction cut-off
- A single, common set of components and component properties

The compositional description of each fluid to be handled in the model should be entered so that it can use a fully compositional EOS to represent fluid phase behavior. For this purpose, each fluid is defined in terms of components that have static properties (e.g., T_c, P_c, Z_c) and a composition (i.e., mole fractions of each component). A section of the component summary is shown in **Fig. 3-6.**

Name	Mnemoic	Mol weight	Crit T.	Crit P.	Crit V.	Crit Z.	BP	A factor	Ref density	.
			degF	psig	ft³/lb mol		degF		lb/ft³	
Water	H_2O	18.00	706.73	3190.64	0.9179	0.2290	211.73	0.3440	62.3031	.
Nitrogen	N_2	28.01	-232.51	476.99	1.4385	0.2917	-320.35	0.0390	50.1921	.
Carbon Dioxide	CO_2	44.01	87.71	1055.69	1.5041	0.2744	-63.67	0.2390	51.5700	.
Methane	C_1	16.04	-116.95	652.48	1.5890	0.2862	-258.79	0.0110	26.5319	.
Ethane	C_2	30.07	90.05	693.09	2.3755	0.2793	-127.39	0.0990	34.2105	.

Fig. 3–6. Compositional component summary

Even if a few of the individual fluid models are of the black oil-type, these can be mapped into the common compositional description. Black oil fluid models can include water and volatile inorganic species such as N_2, CO_2, and H_2S as mole percentages if they are to be included in a compositional description.

The compositional library in the software should hold the characteristics of sufficient hydrocarbon and inorganic compounds. These descriptions are used in the Peng Robinson, Soave Redlich Kwong, Patel Teja, or Valderrama–Patel–Teja EOSs to calculate accurate phase splits and fluid properties throughout the network. Tuned EOSs derived from other sources can be used where available.

For compositional modeling, the user defines the fluid models (dry gas, gas condensate, and oil) specifying the fluid components and their molar fractions and chooses the EOS that is suitable (**Fig. 3-7**).

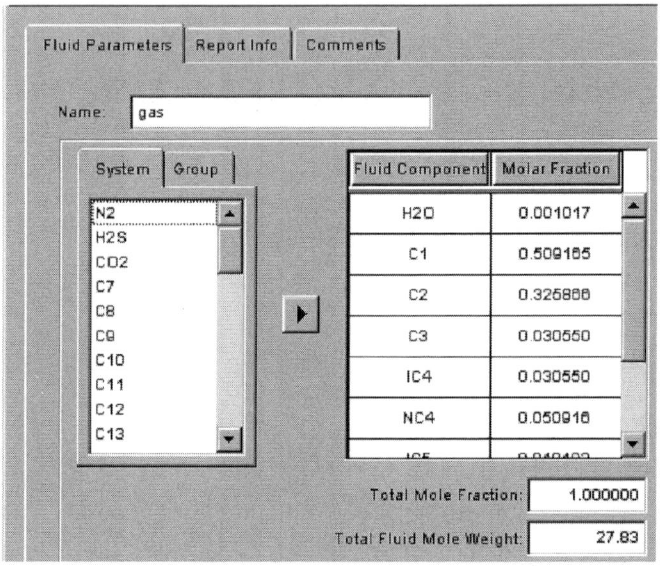

Fig. 3–7. Compositional model for gas

If a compositional PVT report has 25 carbon items (C1–C25), it may not be necessary to use all 25 values. A *Cn+ plus fraction* that is composed of all components of carbon number *n* (cutoff) and higher can be created and used. All of the compositional fluid models within an FSG are rationalized onto a common fluid description (CFD), with a single common plus fraction.

The *plus fraction cutoff* and the number of *pseudo-components* that would be used to represent the group plus fraction in the CFD should be specified by the user.

The pseudo-components are "fictitious" species, whose collective properties can be used to represent the PVT behavior of all the individual fluid models' plus fractions.

The number of pseudo-components required for a representative CFD depends upon the spread of molecular weights of the individual fluid models' plus fractions.

Suppose Fluid A contains only C1 and C2, and Fluid B contains C1...C6 and C7+. A CFD can be created using a *group plus fraction of* 6 and *plus fraction split of 2*. To achieve this, additional pseudo-components should be added to Fluid A, as shown **Table 4-2**.

Table 3–2. Common fluid description for a compositional FSG

Raw data		Data with pseudo-components		FSG Description
Fluid A	***Fluid B***	***Fluid A***	***Fluid B***	***CFD***
C1	C1	C1	C1	C1
C2	C2	C2	C2	C2
	C3	C3 (0)	C3	C3
	C4	C4 (0)	C4	C4
	C5	C5 (0)	C5	C5
	C6	C6 (0)+	C6+	C6+ → C6-1, C6-2
	C7+			

Once the FSG has been merged onto the CFD, it can be assigned to the current snapshot, and this will be used for the optimization calculations.

The accuracy with which the software will predict the behavior of fluids in the models depends upon the quality of the PVT data and how well the chosen EOS "fits" the observed PVT data. The latter can be improved by tuning the EOS to the observed data.

A compositional FSG is constructed and used in a network model as shown in **Fig. 3-8**.

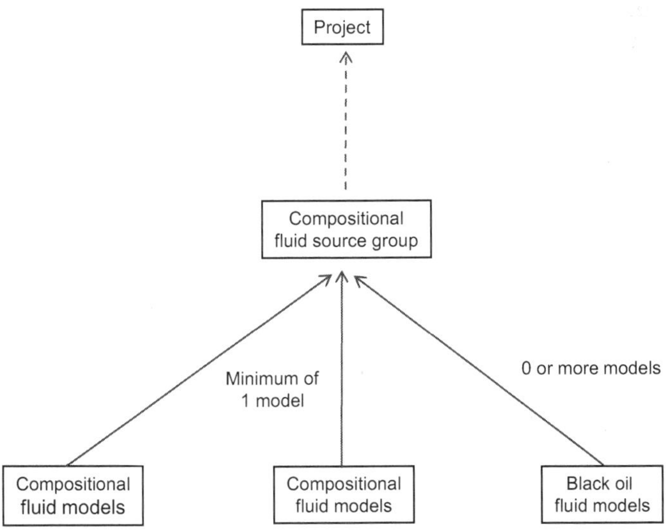

Fig. 3–8. Compositional FSG construction

3.3 Incorporating Wells in a Network Model

Wells are already modeled using the nodal analysis software. Accessing the calibrated well model from the well performance curve generation window of the network model, the well performance curve should be generated against a range of flowing wellhead pressures at the surface reference depth, and included in the network.

Upon receiving the command to generate the well performance curve, the network modeling software will internally launch the nodal analysis software, use the assigned well model with its PVT and generate the performance curve, and attach to the wellhead object. Each wellhead object, therefore, will always have a performance curve attached to it. This approach leads to faster optimization and is better than dynamically linking the well models. The model is up-to-date and ready for any run at any time.

With every recalibration of the well model, a new performance curve will be generated, and the new performance curve will replace the old one. The same approach is followed for naturally flowing (NF) as well as artificially lifted (AL) wells. The fluid model that was used for well modeling based on the formation zone of production will be assigned to the wellhead object in the network model.

3.3.1 Wellhead Object

The Christmas tree represents the wellhead object to be connected at the surface. The wellhead choke and block valves are modeled as a part of the surface network, downstream of the wellhead for producers.

The wellhead objects for various types of wells are differentiated in visual appearance so that the type of well is easily visible to any user (N-NF well, G-GL well, and E-ESP well), as shown in **Fig. 3-9**. They are also color-coded differently. The process connections for connecting the wellheads are also shown with the objects.

Both the N-well and E-wells have outlet socket points.

The G-well has an inlet socket for gas lift injection and an outlet socket for fluid production.

An injector well will have an inlet socket to accept injection fluids.

3.3.2 Naturally Flowing Wells

The well performance curve of an NF well is a graph of the oil production rate plotted against the surface wellhead flowing pressure. It is a simple two-dimensional curve with Q_{oil} plotted on the abscissa against FWHP plotted on the ordinate.

A wellhead performance table (WPT) has the performance data of wellhead pressure and temperature versus flow rate. The WPT data can be input manually, or cut and pasted from Excel or similar applications, or imported automatically from the most recent well model.

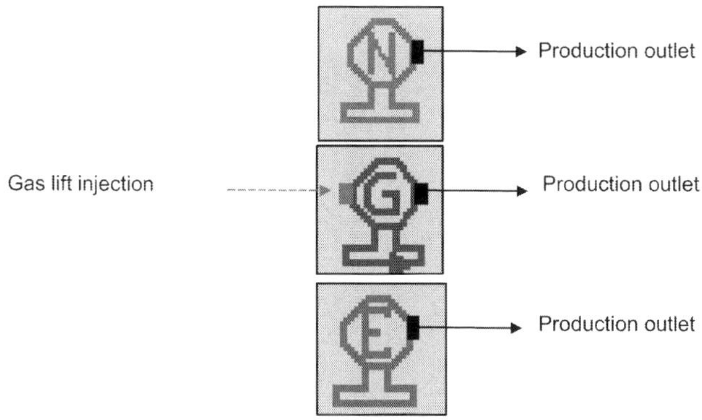

Fig. 3–9. Wellhead objects display

The composition and component properties of the fluid flowing from the well are defined by the fluid model assigned to the wellhead object.

The range of pressures for which the performance curve should be generated is specified by the user. The well performance curve is usually generated for a broad range of FWHP values around the existing value. This will ensure that data of sufficient bandwidth is available for interpolation on either side during optimization iterations.

A sample performance curve for an NF well is shown in **Fig. 3-10**. The pressure and temperature curves are plotted on the primary and secondary vertical axes against the oil production rate plotted on the horizontal axis. If the performance curve is viewed in a network model with a converged solution, the operating points (OPs) are also marked on the graphs.

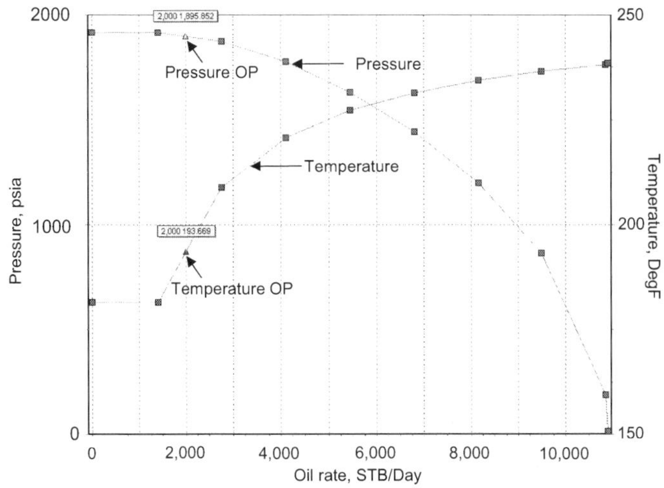

Fig. 3–10. Performance curve of NF oil well

A wellhead object functions as a source of fluid for the network. Each fluid phase at the wellhead (gas, oil, or water) can have revenue (price) or cost associated with its production. This information is used in conjunction with other objectives and constraints to define the optimization problem, for both black oil and compositional descriptions (to be discussed in detail later, in **Section 4.2** on *Objectives and Constraints*).

3.3.3 Gas Producer Well

A gas producer will have similar performance but will show the gas flow rate versus the surface wellhead flowing pressure (**Fig. 3-11**).

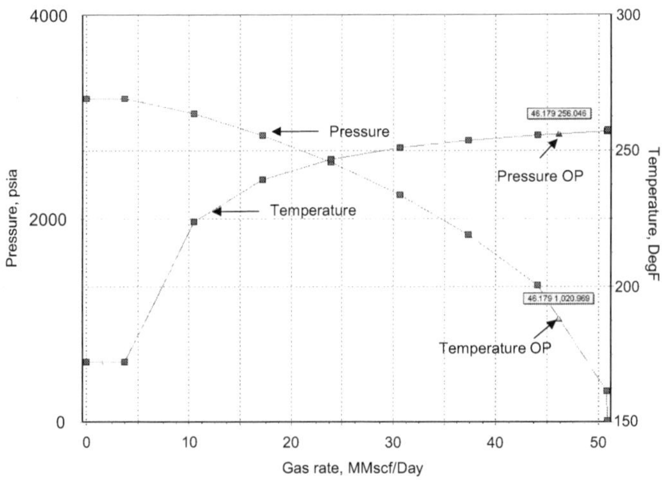

Fig. 3–11. Performance curve of gas producer well

3.3.4 Gas Lifted Wells

The performance curve for a gas lifted well directly obtained from well modeling software is the gas lift response curve. This is a graph of the oil production rate plotted against the gas lift injection rate for a specific FWHP.

In a network operating environment, the surface flowing pressure of the well will be determined by considering the interaction from other wells, as a part of the network optimization. Therefore, the well performance curves of a gas lifted well in a network model should be available for a range of FWHP values. Such a presentation of a family of well performance curves is called the *well performance surface* (WPS). This allows the optimizer during iterations to work in a three-dimensional space (interpolate between the performance curves for different FWHP values) and identify the oil production rate for a gas lift injection rate at a FWHP determined during the network optimization.

Using gas lift injection pressure and rate available for the well, the well modeling software generates a performance curve for the range of wellhead pressures. The wellhead flowing pressures specified in the model should cover the expected range of results so that interpolation will be more accurate. As an example, if the expected FWHP falls between 200 and 250 psig, curves may be selected for pressures 100–400 psig in the intervals of 50 psi (7 curves).

Keeping the variables used for the curve generation in the background, a simple performance curve is plotted between Q_{oil} and Q_{gi} for every FWHP with necessary cut-off. The WPS of a high water cut well that cannot flow naturally is shown in **Fig. 3-12**.

The WPS is generated assuming a single-point lift in the well. It is assumed that the unloading valves (if any) in the mandrels above the operating valve are closed. The software treats all mandrels other than the designed operating point as *inactive* mandrels.

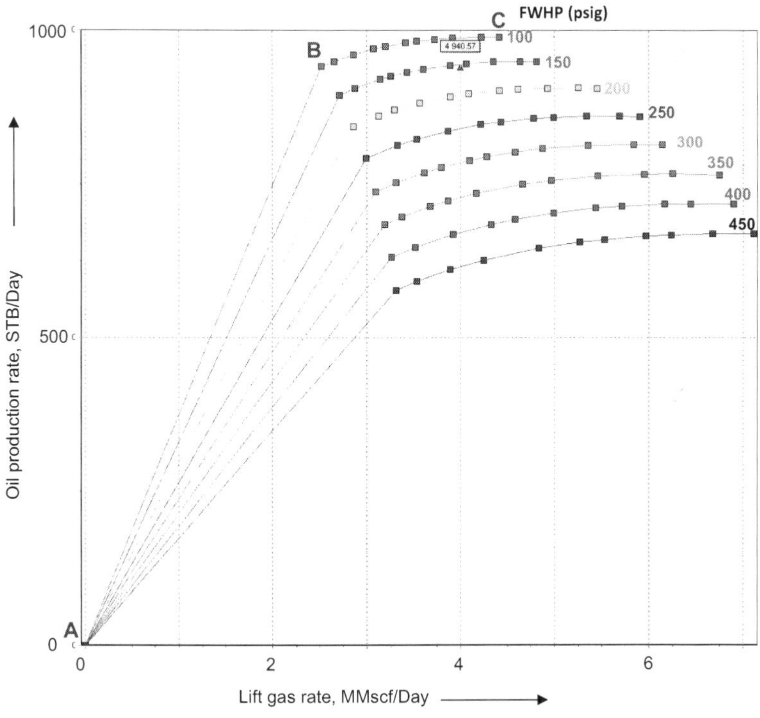

Fig. 3–12. Gas lifted well WPS—low pressure

In the WPS shown above, the curve ABC represents the performance curve for an FWHP of 100 psig. At an FWHP of 100 psig, the well requires a minimum lift gas rate of 2.5 MMscf/Day for gas lift operation through the designed operating depth and will produce 940 STB/Day of oil with the current water cut conditions.

Even though points A and B are valid points, the section AB is an *invalid section of the curve*—this means that if the gas lift allocated is, say, 1.5 MMscf/Day, the well will have no production. In other words, the network model will treat 2.5 MMscf/Day as the minimum gas lift rate ($Q_{gi\,min}$) in this case; if the program cannot allocate 2.5 MMscf/Day, it will permit only 0 MMscf/Day (the well will not be placed on gas lift), and well production will be nil.

There are wells that can flow naturally at a lower FWHP, but will require gas lift to produce at higher FWHP values. The WPS of a well from a high-pressure reservoir is shown in **Fig. 3-13**. The curves have been generated for a wide range of 200–1800 psig, with increments of 200 psi for demonstration.

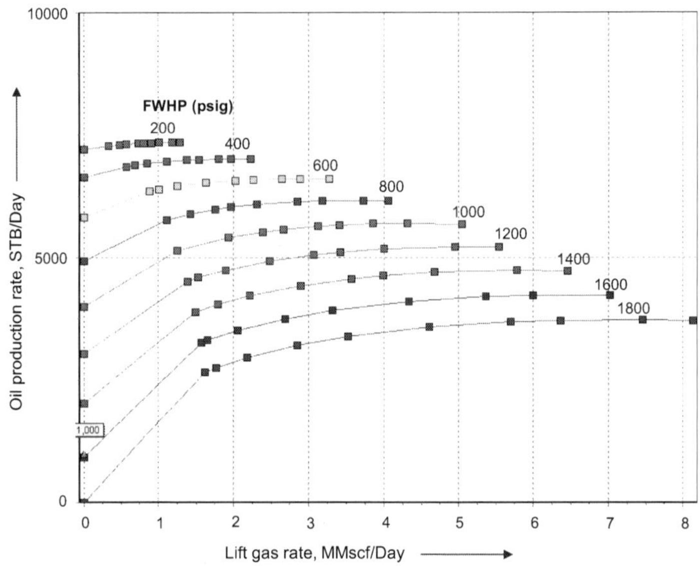

Fig. 3–13. Gas lifted well WPS—high pressure

As can be seen, this well has a very high reservoir pressure and very low water cut—it can flow naturally with an FWHP of up to 1600 psig. This means the lift gas pressure should be more than 2000 psig, at least. The well still requires a $Q_{gi\,min}$ for every FWHP to flow with gas lift in a stable manner. For example the well can flow in a stable manner with a lift gas rate of 0.75 MMscf/Day for an FWHP of 400 psig, but will require 1.50 MMscf/Day for a stable gas lift with an FWHP of 1400 psig.

It is important to recognize that with an increase of tubing pressure, the surface gas lift injection pressure required to keep the well on gas lift also increases. So, if sufficient casing pressure is not available or if the gas lift valve cannot admit the desired quantity of gas, gas lift will not work in the well as designed. Thus, when generating the performance curve for a gas lifted well for the range of FWHP values prescribed, the model should take into consideration the variables on the gas lift side—the available gas lift injection pressure and the gas passage performance of the operating valve—apart from the well inflow-outflow performance and PVT. The gas passage performance of the operating valve depends on upstream and downstream pressures and port size of the valve.

For gas lifted wells, the network modeling software should integrate the true gas lift valve performance and model the gas lift side along with the production side. The assumption that any amount of lift gas rate is available / can be injected for gas lifting the well is invalid. The gas lift pressure available in the network is controlled by the lift gas supply available (a source of gas supply with a defined range of pressures, or a compressor with its prescribed horsepower or performance curve). The software also calculates the pressure drop in the gas lift supply pipeline for the lift gas flow rate to estimate the gas lift injection pressure at wellhead. Knowing the port size of the gas lift valve installed at the operating mandrel, the gas passage performance of the operating valve is also known. With this information, the performance of a gas lifted well is defined as a four-dimensional curve, as shown in **Fig. 3-14**.

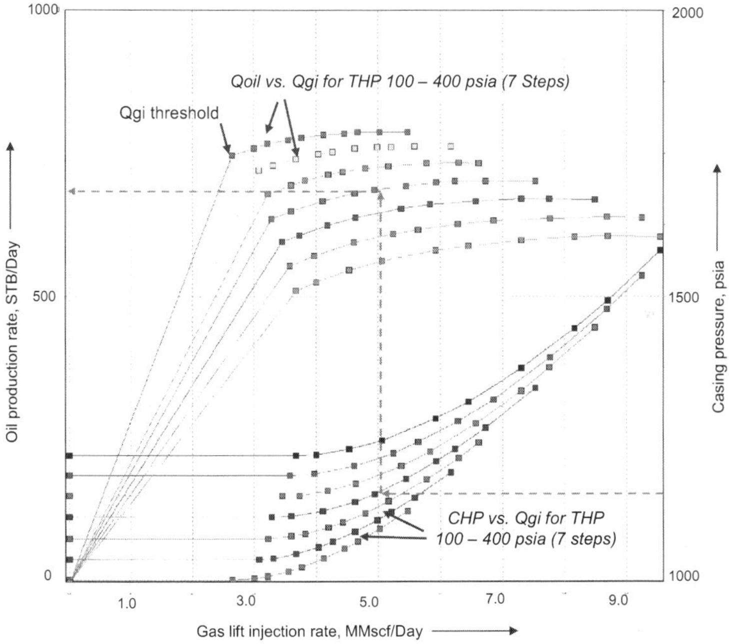

Fig. 3–14. Gas lift injection-production interaction

As can be observed from the plot, if the wellhead pressure is 250 psia, for an available gas lift pressure of 1150 psia, with the existing operating gas lift valve, the well will inject a lift gas rate of 5.0 MMscf/Day and will produce at about 700 BOPD (follow the dotted line).

3.3.5 Pumped Wells

Most of the large production networks may have GL and ESP lifted wells along with the NF wells. For the ESP wells in the network, the downhole pump data is already included in the well model. The upper and lower flow limits as per the manufacturer's recommendations are also defined in the well model. If needed, the ESP "operating envelope" can be input as constraints in the network model. These bounds can be imported from the well model or user-provided.

The wellhead performance table (WPT) for a pumped well that relates wellhead pressure and temperature to flow rate and pump power requirements at the fixed operating frequency is shown in **Fig. 3-15**.

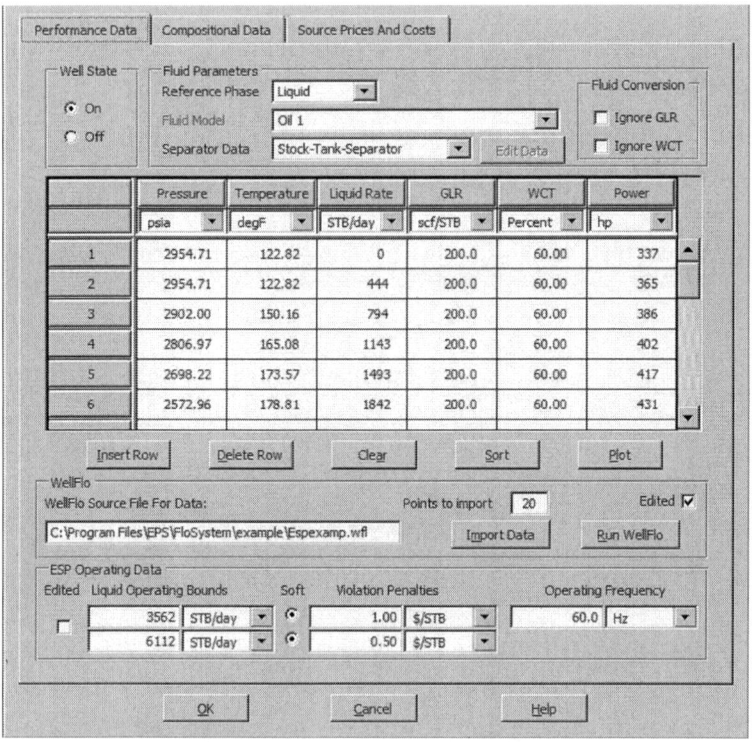

Fig. 3–15. Wellhead performance table of ESP well

The data from the WPT in **Fig. 3-15** can be plotted to produce the well performance curve for the ESP lifted well.

3.3.6 Water Injectors

The performance curve of a gas or water injection wellhead object contains the injection flow rate against the surface injection pressure.

Depending on the water injection supply pressure available at the well location, the range of the pressures will be input in the performance curve generation panel. The injection fluid PVT will be assigned to the wellhead. Using the injection well model already constructed, the surface performance curve will be generated.

The producer well adds fluid into the network while the injector well removes fluid. A positive sign is used for the produced fluids, and a negative sign is used for injected fluids in the network model.

As per the performance plot shown in **Fig. 3-16,** the water injection well can inject 7000 BWPD at a surface injection pressure of 1600 psig.

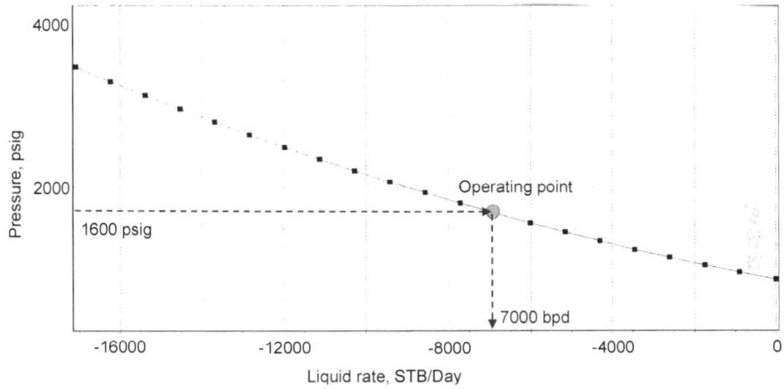

Fig. 3–16. Water injection well performance curve

3.3.7 Gas Injector Well

The well performance curve of a gas injection well is shown in **Fig. 3-17**.

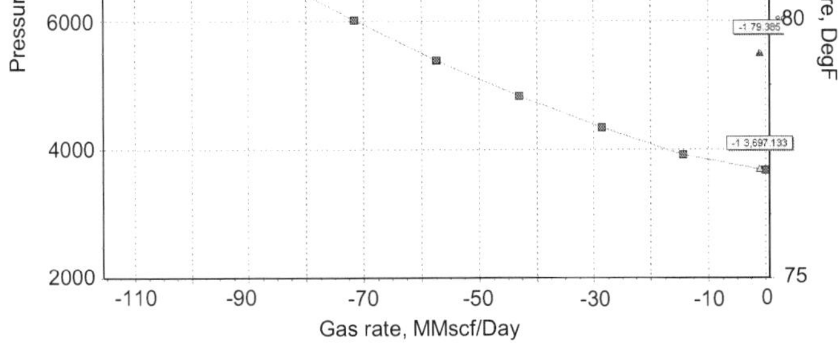

Fig. 3–17. Gas injection well performance curve

3.4 Model Construction Objects

3.4.1 Toolbar for Model Construction

There are several ways to construct a network model. Given below is a methodology commonly used in the modeling industry.

The general toolbar has standard functions along with the Pipes, Links, Optimization Tools, and Network Tools, as shown in **Fig. 3-18**.

The pipe and link objects can be dragged and dropped in the screen while building the network. These construction tools get disabled once the network model construction is over.

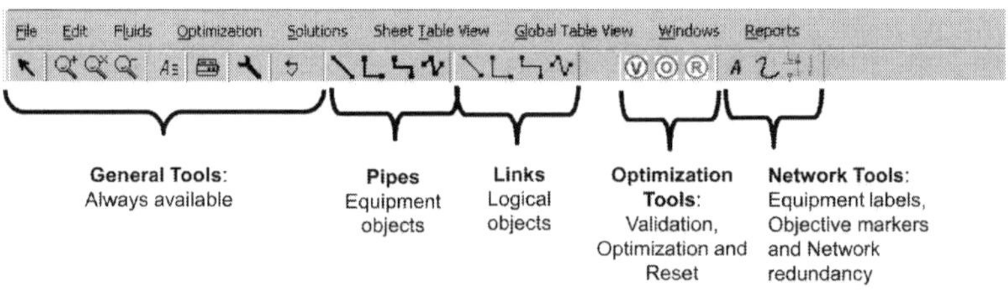

Fig. 3–18. General model construction tool bar

Shown below is a view of an object tool bar that displays some of the common "objects" that will be used to build a network (**Fig. 3-19**). The names are also included alongside the objects for clarity. There are more objects needed for some networks that are available (not included in this list).

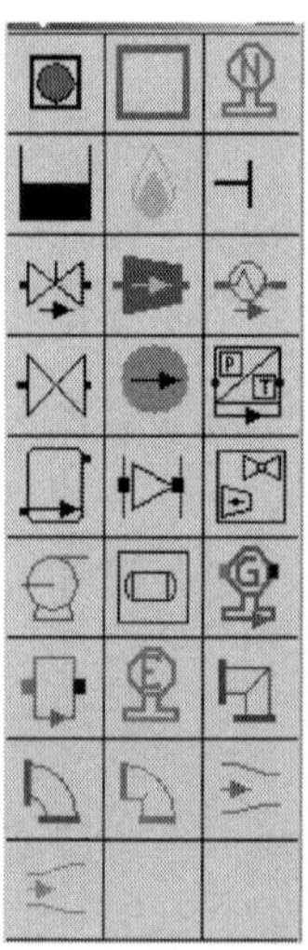

Junction	Sub-sheet	Wellhead (NF Well)
Sink	Flare	Stopper
Choke	Simple compressor	Heat exchanger
Block valve	Source object	Tuning buffer
Gas purifier	Flow direction	Detailed compressor
Pump	Detailed separator	Gas lifted wellhead
Flow group	ESP well	Mitre bend
Standard bend	Standard elbow	Contraction
Enlargement		

Fig. 3–19. Model object tool bar

3.4.2 Pipeline Modeling

3.4.2.1 Scope of Pipeline Modeling

Pipelines are used very frequently when building network models. A simple network model may have hundreds of pipelines. A medium-sized network model may have thousands of pipelines. As the fluids flow through the pipelines, the model should be able to calculate the pressure drop in the pipeline and also the heat transfer from / to the surrounding medium. For every pipeline, starting with the inlet pressure, temperature, flow rate, and fluid properties, the model will determine the pressure and temperature and *in-situ* fluid properties all along the pipeline and at the arrival end.

The following details are required for every pipeline in the model. The *pipe properties* data section is divided into four tabs (**Fig. 3-20**):

End 1 Connected To		Choke_1	
End 2 Connected To		Junction_0	
Dimensions ●	Flow & Tuning ○	Background ○	Erosional Velocity ○
Pipe			
Measured Length		1000	ft.
Absolute Roughness		0.0012	in.
Outside diameter		4.5	in.
Weight/Unit Length		11.33	lb/ft
Inside diameter		4.0	in.
Weight/Unit Length ○		Inner Diameter ●	
Material		Carbon Steel (AISI 1010)	
Density		488.936	lb/ft3
Thermal Conductivity		36.92	BTU/(hr ft DegF)
Geometry			
Elevation Angle		0.00	Degrees
End 1 Elevation		0.00	ft
End 2 Elevation		0.00	ft
Angle ●		Elevation ○	
OK	***Close***	***Import***	***Help***

Fig. 3–20. Pipe data entry tabs

- *Dimensions*: Pipe properties and geometry. Pipeline details such as OD, ID, measured length, absolute roughness, and material (carbon steel, stainless steel and grade, alloy steel, PVC, etc.) are entered or selected from the drop-down menu.

  Geometry: Elevation at the two ends of the pipeline (or elevation angle). This defines the pipe inclination.
- *Flow & Tuning*: User selects the flow correlation to be used for the pressure drop calculations.
- *Background*: Details on insulation, background, heat transfer, and environment are input here. The user enters the insulation material and thickness for every pipe. The background temperature is also entered (air, water, or earth, depending on whether the pipeline is on the surface, subsea, or buried in the earth). Insulation options such as air, polyurethane foam, clay, concrete, and water are available. Most of the standard information about piping can be copied down and filled by a single click for any number of pipelines.
- *Erosional Velocity*: Erosional velocity check is performed and flagged for every convergence of the model. The erosional velocity (V_e) is calculated using the formula:

$$V_e = \frac{C}{\sqrt{\rho_f}}$$

  Where ρ_f is fluid density (lbs/ft^3). The default value of constant C (in field units) is 100 (somewhat conservative), but can be changed by the user.

The pipeline toolbar contains four buttons (**Fig. 3-21**) that control the display only; they have no physical effects on the pipe behavior or calculations. These are used for model construction.

	Draws a pipe which is displayed as a straight line between objects
	Draws a pipe with two segments, one horizontal and one vertical, with one 90° bend
	Draws a pipeline in three segments, with two 90° bends
	Draws a pipe using single mouse-clicks to define multiple segments between two objects. Double-click terminates the pipe.

Fig. 3–21. Pipe appearance

The ability to model pipelines, flowlines, and risers includes:

- Industry-standard single and multi-phase flow correlations
- Fully coupled pressure temperature modeling

- Emulsion viscosities
- Advanced environmental modeling and heat transfer
- Erosional limit flags

Normally the term *flowline* is used to represent the short section of a pipeline from the wellhead to the production manifold. The term *pipeline* is used to refer to the long pipelines, e.g., a pipeline carrying produced fluids from the production manifold to the separation facilities or a gas supply line from the gas compressor to the well platform. The *riser* or *downcomer* is a vertical pipeline normally used in offshore and subsea applications. All of these are included in the model.

3.4.2.2 Pipeline ID and OD Data

Oil companies use pipe that conforms to certain international standards specifying the dimensions and weights of welded and seamless pipe. The relevant standards are ASTM International (American Society for Testing and Materials) or ASME (American Society of Mechanical Engineers) or others. The spec also covers the chemical composition and impurities allowed in the pipe manufacture.

Ready reckoners to specify the dimensions and weights of welded and seamless pipe are available from manufacturers. Knowing the nominal pipe size (NPS) and the class of the pipe, OD and ID of the pipe (wall thickness) are known. The class of the pipe is usually mentioned as Schedule 10 through 160, STD, XH and XXH, and so on. Data for pipes with NPS ¼" to 30" diameter has been published by manufacturers and agencies and are readily available in the public domain for reference.

Piping engineers select the grade, class, or schedule of the pipe based on the service it is intended for, and they ensure that the pressure rating is adequate for the service.

A few samples of data of pipelines normally used for gas lifted wells and platforms are shown in **Table 3-3.**

Table 3–3. Sample pipeline dimensions

Nominal size (in.)	Schedule	OD (in.)	Wall thickness (in.)	ID (in.)	Pressure rating (psig) at 200°F [Est.]
2	80	2.375	0.218	1.939	2369
8	160	8.625	0.906	6.813	2738
6	80	6.625	0.432	5.761	1650
24	80	24	1.219	21.562	1266

3.4.2.3 Level of Detail Required for Pipeline Modeling

In a field, thousands of pipelines are in service. The question arises whether the physical dimensions of all the pipelines should be included in the model. The following discussion is intended to clarify this issue.

Considering the modeling of production flowing from a well to the production separator, the model will require the physical properties of the pipeline to accurately model the loss of pressure and heat.

Fig. 3-22 shows the field production lineup of a well to the separator.

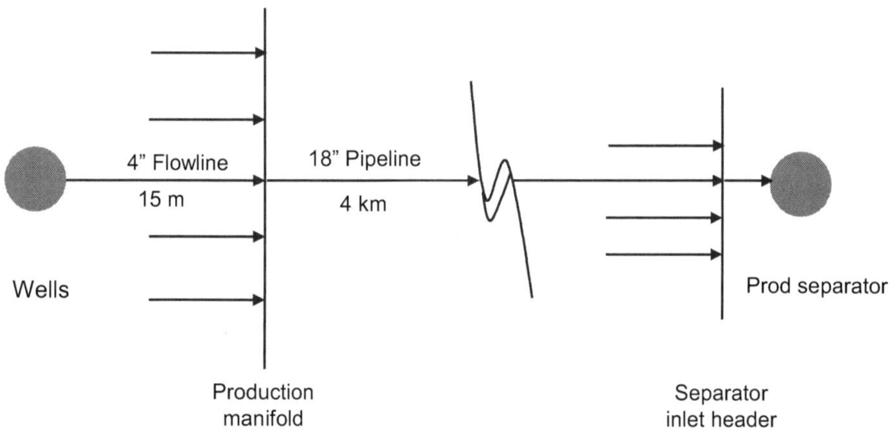

Fig. 3–22. Production line modeling details

When the flowline connected to the Christmas tree is 15 m long, the departing pipeline from the production manifold is 4 km long. It would appear that the pressure drop in the flowline may be insignificant due to its short length. However, this cannot be generalized. It would really depend on the size of the flowline (4" in this case), current condition of the pipe, and the flow rate from the well. The field pressure data will clarify if the flowline segment is a constraint. Generally, it is a good practice to model the full details of each flowline, as it would not be known at the time of modeling whether the flowline is a bottleneck or not.

3.4.2.4 Pipe Fitting Modeling

To handle the pressure drop caused by pipe bends and other fittings, some engineers add 10% length to the actual length of piping in the field and consider it sufficient to account for the pressure drop in the bends and other fittings. The accuracy required depends on the optimization objectives and well characteristics. If the well productivity index (PI) is 10 BPD/psi, every PSI of pressure loss becomes significant, but not so for a well with PI of 0.5 BPD/psi.

For pressure drop calculations to be accurate, details of various pipe fittings such as elbows, bends, pipe expansion and constriction, check valves, and chokes should be included in the model. Similar details for other wells can be filled in using bulk data entry/edit methods.

The pressure drop in pipe fittings is included as *equivalent lengths* of straight pipe. In other words, the equivalent length of the pipe fitting would be added to the length of the pipeline.

The software would calculate and apply equivalent lengths for all pipe fittings used in the field. The equivalent length calculations are usually based on methods described by Crane.[17]

The following illustration demonstrates how the pipe fittings add up to the length of the flowline in pressure drop calculations. **Fig. 3-23** shows the details of a 6″ oil production flowline from a producer well Christmas tree, connected to the production manifold.

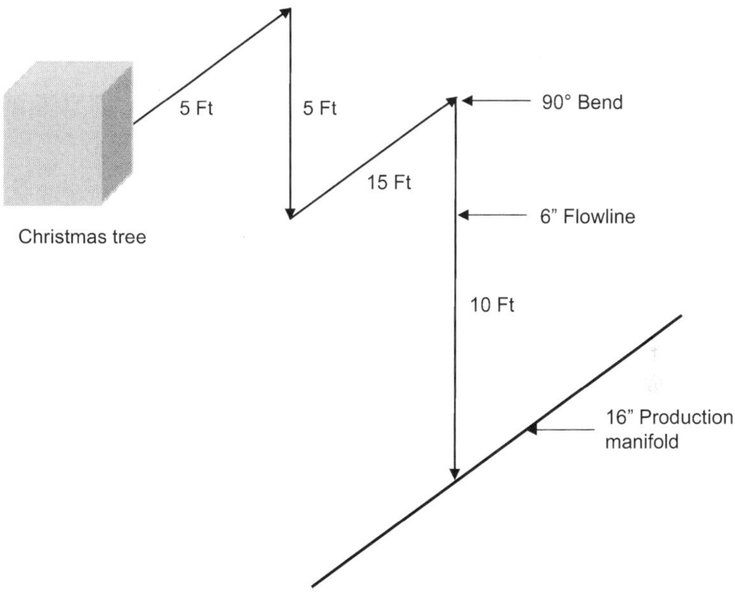

Fig. 3–23. Effective flowline length

The flowline has three 90-degree elbows before it joins the production manifold. Thus, the equivalent lengths of the elbows would be added to the length of the flowline piping.

For a 90° elbow with an r/d ratio of 2, the $L_e/D = 12$

The equivalent length for a 90° elbow, $L_e = 12 \times 6$ in $= 6$ ft.

The effective length of flowline = $5 + 5 + 15 + 10 + 3 \times (6) = 53$ ft.

Thus, the effective length of the flowline becomes 53 ft.

Some common pipe fittings—standard bend, mitre bend, standard elbow, contraction, and enlargement are included in the object tool bar (**Fig. 3-19**).

3.4.2.5 Pressure Gradient in Horizontal Pipelines

The network model calculates several *in-situ* parameters for fluid flow in pipelines. After a model run, these parameters of the major pipelines in the model that carry production from wells to separation area can be compared.

Among horizontal pipelines, one quick method to find out which multi-phase pipeline is heavily loaded is to simply review the pressure gradient. As the pipe is horizontal, the pressure drop is due to friction only.

Pipeline identification	Pipe size, in.	Length of pipe, Ft	Inlet/outlet pressure, psig	Pressure gradient, psi/ft
Pipeline 1	16	5000	235 / 135	**0.02**
Pipeline 2	18	12,500	275 / 135	**0.0112**

Pipelines 1 and 2 have a pressure drop of 100 and 140 psi, respectively. However, the pressure gradient in Pipeline 1 is more than Pipeline 2. Reviewing the flow rates, the network model can be used to evaluate if the greater frictional loss is due to pipeline plugging or fluid loading. Corrective actions should be taken, wherever possible, to improve the service life of the pipelines.

3.4.3 Use of Links in Place of Pipelines

Pipes are physical connections. *Links* are logical connections. Modeling engineers use links in some locations in the model instead of pipes, where the pressure drop is insignificant or it does not matter for the optimization. Links allow fluid flow, but do not have any pressure drop or heat transfer. There is no representation for the diameter or length of the flow path.

A wing valve is bolted directly to the body flange of the Christmas tree. A link can be used in the model to connect the wing valve to the wellhead object.

The appearance of links will be different from pipes so that the user can distinguish them visually. Pipelines are continuous lines. Links are dotted lines. The color scheme used for pipelines can be used for links also to display the type of fluid being carried.

A link used in a model can be replaced with pipe later, if needed, when the data becomes available or the modeling of pressure and heat loss becomes important.

3.4.4 Selection of Flow Correlations

The system optimization program uses flow correlations to calculate pressure drop and heat transfer in pipes. The user selects one of the available flow correlations based on the type of fluid that runs in the pipe. The user also selects appropriate heat transfer coefficients. Modeling companies keep adding correlations in their software based on suitability, industry requirement, and feedback. Much research has been done in the industry on this subject. Some of the standard correlations are mentioned below:

Gas flow correlations

- Panhandle_A
- Panhandle_B
- IGT
- Weymouth
- Average Velocity Model
- Integrated Gas Correlation

Multiphase flow correlations

- Beggs and Brill
- Modified Beggs and Brill
- No-slip Beggs and Brill
- Dukler-Eaton-Flanigan

The pressure drop calculated by the correlations may not match perfectly with the pressure drop measured in the field. The pressure drop and heat transfer along a pipe can be fine-tuned by small adjustments to the **L**ength or **D**iameter or Heat Transfer (**U**). These factors act as multipliers on the pipe variables already entered. These should be used with caution. These can be used during history matching to compare how the actual pressure drop differs from the theoretical pressure drop estimated by the correlation.

Some models offer an emulsion modeling option that allows the user to account for the effects of emulsion formation on the fluid viscosity that would affect the pressure drop in the pipe. This may be used if emulsion is observed in produced fluids.

Apart from the pressure and temperature change, the program has the ability to calculate several additional results for every successful run of the model. A few results of interest are indicated below:

- Flow rates at *in-situ* and stock tank conditions in every pipeline
- The mixture velocity in the pipe and superficial velocity of gas and liquid phases
- *In-situ* liquid hold-up / gas-void fraction and slippage (if present) in multi-phase flow pipelines
- Conditional check if erosional velocity has been reached

- The overall pressure drop in a pipeline (also reported in the form of pressure gradient)
- Any other flow assurance issues identified

3.4.5 Modeling of Block Valves

A *block valve* changes the configuration of a network by being either fully open or shut. The valve can be used to shut in a well or isolate a section of a network.

The block valve has one inlet and one outlet socket point, but is not directional. The type of valve is immaterial—gate valve, globe valve, ball valve, etc.

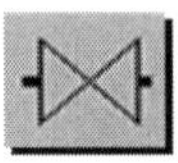

Open Block Valve

When the block valve is open, there is no pressure drop across it.

Closed Block Valve

When the block valve is shut, there is no flow through it.

In an offline optimization process, the user sets the valve in either open or closed position. In an online optimization, based on the settings in the model, the optimizer may open a valve and admit flow through it, based on the model requirement. This would mean that the valve is remotely operable in the field (actuated by an electrical, pneumatic, or hydraulic actuator).

The block valves are very useful in isolating a part of a network that need not be used for optimization. For example, if a condensate recovery unit (CRU) is out of operation, it should be bypassed in the model so that it can be ignored during optimization. The bypass valve should be opened and the inlet and exit valves of the process unit should be closed, as shown in **Fig. 3-24**.

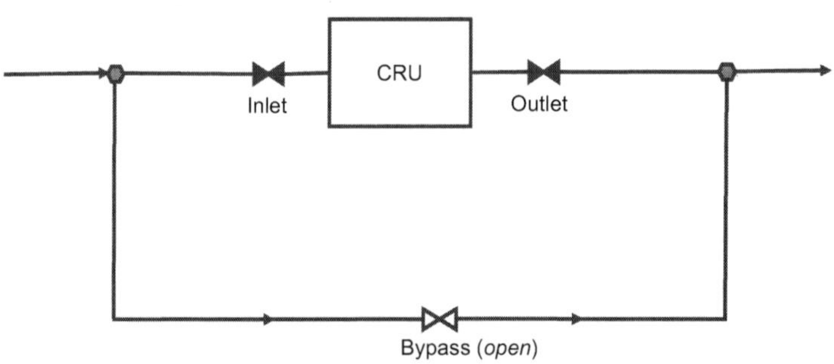

Fig. 3–24. Use of block valves to isolate part of a network

Fig. 3-25 shows a view of a simplified production platform where the wellhead chokes are on optimizer control. Each well can be lined up to one of the three headers—HP header, LP header, or the test header—using respective block valves.

Complex fluid routing to production manifolds can be modeled by explicitly providing all possible tie-in options physically available in the field, and then closing off the options that are currently inactive by using the block valve object. The presence of arrows indicates that the model has converged and has a solution.

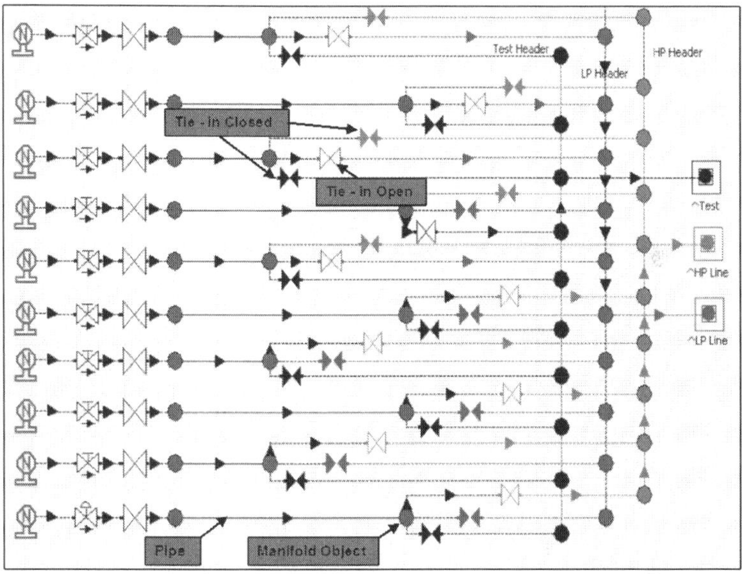

Fig. 3–25. Use of block valves for diverting production

3.4.6 Modeling of Chokes

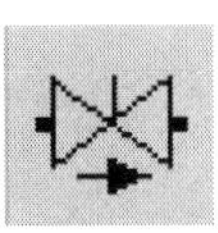

Chokes are very important parts of any network. They control or regulate the flow. The choke object allows a pressure drop to be modeled at a single point in the network. A choke has one inlet and one outlet and is a *directional* object. Chokes can be connected to other items by pipes and/or links.

Different choke models are available for liquids, multi-phase fluids, and dry gas, flowing under different conditions. The user selects the appropriate choke model depending on the role of the choke in the network and the type of fluid. The model results also indicate if the flow through the choke is critical or sub-critical.

There are three different types of chokes: *fixed throat diameter, maximum throat diameter*, and *pressure reducer*. For ease of visualization, the display picture for the choke also changes in the model, based on its function, so that the user knows if the choke opening is controlled by the user or by the optimizer.

A *fixed throat diameter* choke is the same as a choke bean inserted in a flowline of an oil well. The diameter of the choke (restriction) is fixed. The well may have a 6″ flowline. But the flow from the well is controlled by a fixed choke bean with a diameter of, say, 24/64″. The diameter of the fixed throat will be entered by the user. The pressure drop across the choke is dependent on flow rate and fluid properties only.

The *maximum throat diameter* choke is used when the maximum choke trim size is known, but the choke opening is controlled by the optimizer. For example, if an automated choke is used for gas lift injection in a well, the maximum throat diameter is entered in the model. The optimizer can open or throttle the choke during optimization iterations, but the maximum choke opening is limited by the size of trim available in the field. When fully opened by the optimizer, the choke diameter will be the same as the trim size entered by the user.

The pressure drop across the choke is under optimizer control and decided by the rate of flow. The optimizer will adjust the choke opening to satisfy the objectives and constraints set at the choke or elsewhere. The choke diameter associated with the flow rate through, and pressure drop across, the choke is calculated and reported.

The third type of choke is the *pressure reducer*. This is actually a control valve. The model will use the pressure reducer choke to reduce the pressure, but does not attempt to relate the pressure drop across the choke to a choke size.

The pressure reducer mode is useful for modeling pressure control valves (PCVs). In this case, the control valve is used to maintain an upstream or downstream pressure of a vessel, based on the set point input in the model.

3.4.7 Modeling of Compressors

There are two ways of modeling gas compressors in a network: simple compressor and detailed compressor.

The *simple compressor* model relates *compression ratio, horsepower,* and *flowrate.* It does not use compressor performance curves. In other words, the compressor speed is not modeled in a simple compressor model. This model is valid for both centrifugal and reciprocating compressors.

For many applications, the simple compressor object should be sufficient, as it uses the horsepower that is available in the field (**Fig. 3-26**). The user defines the cost associated with the power consumed by the compressor in the model (electric power or fuel gas). As the optimizer tries to maximize the revenue, it will try to optimize the horsepower consumed and reduce the cost.

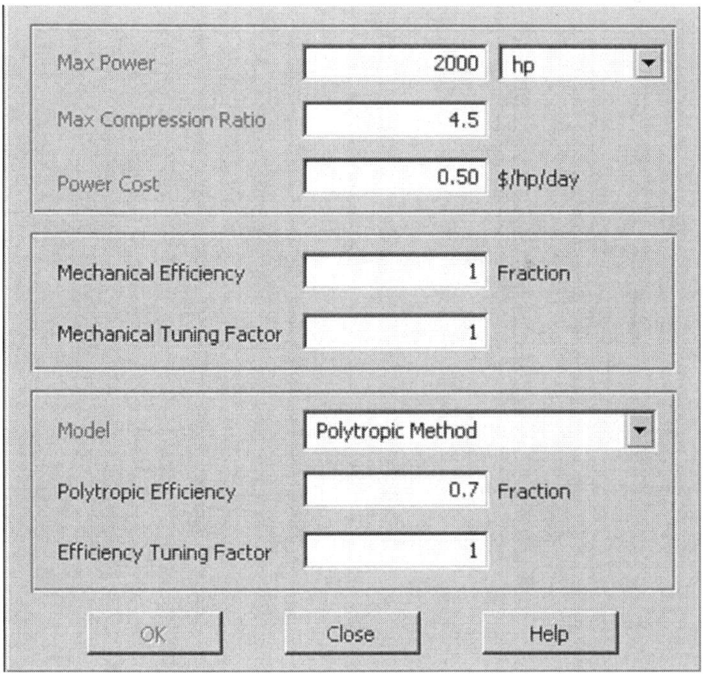

Fig. 3–26. Simple compressor model

When using simple compressors, the user can select either the polytropic or the isentropic process (simple compressors in black oil models are always polytropic). The compressor object has two socket points, allowing one inlet and one outlet connection, and is a directional object. The flow direction through the compressor stage is fixed and is marked by an arrow on the icon. Compressor stages can be connected to other components using pipes and/or links.

Single stage and multi-stage compressors can be modeled using the simple compressor object. **Fig. 3-27** shows a two-stage compressor with accessories such as inter-stage coolers, scrubbers, and anti-surge recycle valves.

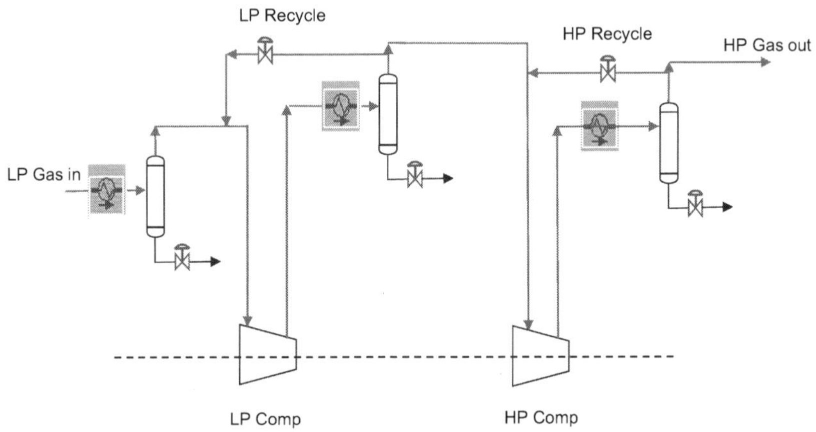

Fig. 3–27. Two-stage compression modeled using simple compressor

For optimizing large gas lifted assets with several giant gas compressors, money spent on compressor maintenance is a significant portion of the operating costs. The detailed compressor is a very comprehensive approach towards compressor modeling. It will model the gas generator, power turbine, gear box, and the multistage compressors connected to the drive with all the accessories. It will incorporate the gas turbine performance and performance curves of every stage of the compressor. This will lead to more accurate optimization of the asset and will bring large benefits to those fields. [18]

The detailed modeling of compressors will be discussed in detail in *Section 3.5*.

3.4.8 Modeling of Pumps

The *pump* object increases the pressure of a fluid. A pump has one inlet and one outlet and is a directional piece of equipment.

If the pump is switched on, the pump model will add energy to the fluid, based on the pump speed, shaft power, and *in-situ* fluid flow rate.

If the pump is switched off, no fluid passes through it, and the effect is the same as a closed block valve (the crossed object shows pump is turned *off*). If the user does not want to stop flow altogether, a bypass loop should be placed across the pump object, so that the flow can continue through the bypass line even if the pump is stopped.

The detailed pump object has pump performance curve and tuning data, apart from switches, bounds, costs, and design constraints for the pump. All data are entered as a function of the *in-situ* flow rate.

There are two pump types:

- **Centrifugal**

- **Reciprocating**

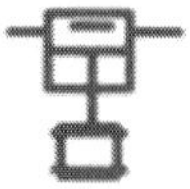

The data required by the two pump types are very similar.

3.4.8.1 Centrifugal Pumps

Centrifugal pumps can pump at a range of flow rates while operating at a fixed speed. The pump performance is usually expressed as curves of the following parameters versus *in-situ* flow rate at a given design speed.

- Head Rise
- Input Power
- Efficiency

Head Rise = $\Delta H = \Delta P/(\rho.g)$, where ρ = density of the fluid and g = acceleration due to gravity

The pump curves are normally produced through testing of the pump, using water as the pumped fluid. If fluid of a different density is pumped, the head versus *in-situ* flow rate curve is still valid, but the power requirement changes in proportion to the change in density. For this reason, the centrifugal pumps are called *constant head pumps.*

Power transferred to fluid = m.g.ΔH = Q.ρ.g.ΔH = Q.ΔP

Where m = mass flow rate and

Q: *in-situ* volumetric flow rate

As an example of a detailed model of a centrifugal pump, the performance of the ESP in a water supply well is given in **Fig. 3-28**. The same data is shown as a set of pump curves in **Fig. 3-29**. Note that the pump curve is for a fixed speed, 3000 rpm.

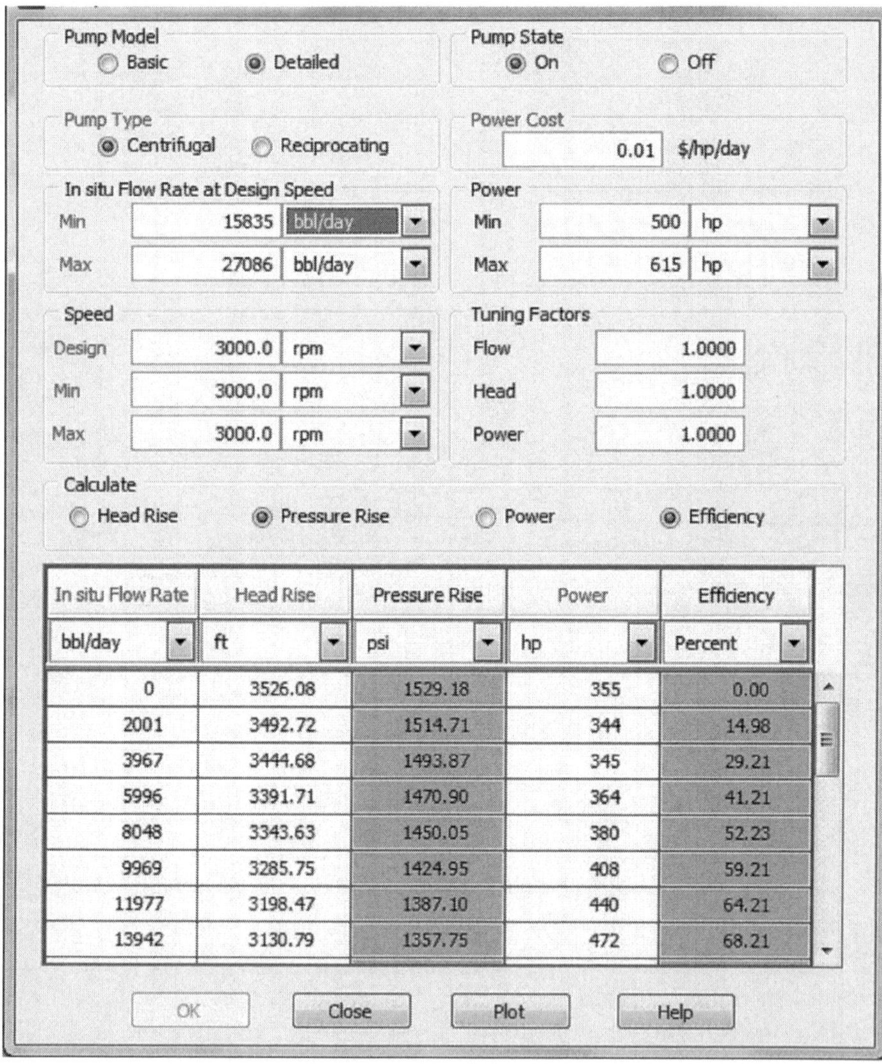

In situ Flow Rate	Head Rise	Pressure Rise	Power	Efficiency
bbl/day	ft	psi	hp	Percent
0	3526.08	1529.18	355	0.00
2001	3492.72	1514.71	344	14.98
3967	3444.68	1493.87	345	29.21
5996	3391.71	1470.90	364	41.21
8048	3343.63	1450.05	380	52.23
9969	3285.75	1424.95	408	59.21
11977	3198.47	1387.10	440	64.21
13942	3130.79	1357.75	472	68.21

Fig. 3–28. Centrifugal pump data

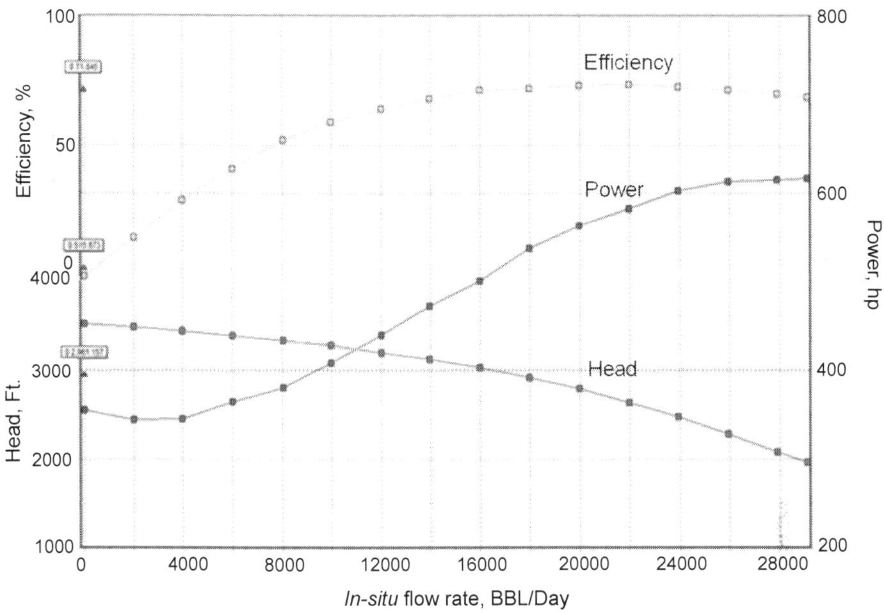

Fig. 3–29. Centrifugal pump performance curve

3.4.8.2 Reciprocating Pumps

For a given speed, reciprocating pumps pump approximately the same *in-situ* flow rate, independent of the head (or pressure) rise. The approximation ignores leakage, which tends to increase slightly with discharge pressure.

At a given flow rate, they can provide any head (subject to the maximum power available and the limits of the machinery). The efficiency and power required are functions of the head rise.

As with centrifugal pumps, the performance curves for reciprocating pumps assume that the pumped fluid is water. If fluid of a different density is pumped, the head versus efficiency relationship at the single *in-situ* flow rate is still valid, but the power requirement changes in proportion to the change in density.

The performance of reciprocating pumps at speeds other than the design speed is given by:

$Q \; \alpha \; (N)$

$\Delta H = \text{Constant}$

$\text{Power} \; \alpha \; (N)$

Note that a reciprocating pump will only have *one* possible operating *in-situ* flow rate unless the speed is allowed to vary.

3.4.9 Modeling of Heat Exchangers

Heat exchangers allow the enthalpy of the fluid flowing through them to increase or decrease, resulting in a temperature change. A pressure drop can also be calculated.

Heat exchangers have one inlet and one outlet, and they are directional equipment items. Heat exchangers can be connected to other items with pipes and/or links.

A basic heat exchanger can be modeled with a target exit temperature of fluid. The heat exchanger can act as a heater only, cooler only, or both.

The *basic* model assumes that the heat exchanger will instantaneously satisfy the heat transfer demand to meet the target exit temperature. Any associated pressure change is a fixed, user-input value.

The *detailed* model calculates the exchange of heat between the process fluid (which flows through the entire network) and a service fluid (which provides the heating/cooling capability and is confined within the heat exchanger object). Any associated pressure drop is rate-dependent and calculated by the model.

3.4.10 Modeling of Separators

In the separator, flash calculations are carried out on individual separator stages to determine the phase split at the operational temperature and pressure. The separated fluids leave the object through different exits. The operational temperature and pressure can be fixed (as constraints) or placed under optimizer control. Multi-stage separation can be modeled by including several separator stages and connecting them appropriately.

Entrainment of gas in oil or water, oil in gas or water, and water in gas or oil can be modeled. This option may be used to indicate that the separation is not fully complete. Note that the coefficients are expressed in terms of mole fraction. The values in **Fig. 3-30** show that 5% of the total gas stream is entrained in the oil stream and 3% of the total oil stream is entrained in the gas stream.

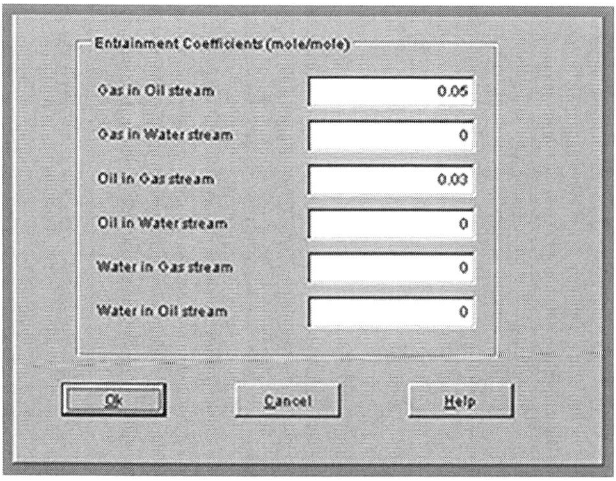

Fig. 3–30. Entrainment coefficient for separator streams

3.4.11 Modeling of Miscellaneous Objects

Sub-sheets are logical, rather than physical, objects used to assist in presentation of the network in the graphical user interface. They have no variables. Each sub-sheet has all the functionality of the parent screen and may contain further sub-sheets. This "vertical stacking" of sub-sheets is similar to that used for file folders in Windows.

Sinks are points in a network where fluid leaves the model. Sinks are generally used to represent delivery points in network descriptions. Apart from objectives and constraints, the sink contains no user-definable parameters that can influence the optimization calculations. Equipment items can be connected to sinks using a pipe and/or a link.

The *flare* object represents a gas flare or vent allowing fluid to leave the network. Functionally, it is equivalent to a sink. Apart from objectives and constraints, the flare does not contain any properties that can influence optimization calculations.

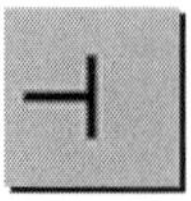

Stoppers are used to close off an open connection in the network, such as

– Pipe ends (direct connection to the stopper)

– Unused equipment outlets (connected to the stopper by a link)

This would represent a spectacle blind in a pipeline in the field.

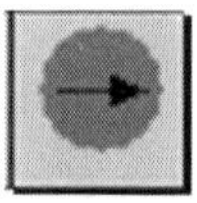

Source objects represent another point of entry of fluid into the system network.

The purpose of this object is to bring a specified fluid into the field within a pre-defined range of rates, pressures, and temperatures. Typical examples of source objects in networks include:

– Fluid from other wells that have not been modeled in this network (e.g., from an adjacent field)

– Introduction of third-party fluid to a process facility, e.g., lift gas, fuel gas, etc.

In a source object, the fluid type, supply pressure, and temperature ranges must be specified. A fluid property should be defined by assigning a proper fluid model. Importantly, the source fluid price or cost should be specified so that the optimizer, while importing the required amount of fluid from the source object, will know the cost it has to pay to bring that fluid into the field (**Fig. 3-31**).

Performance Data ●	Compositional Data ○	Source Prices And Costs ○
Fluid Parameters		
Reference Phase	Oil ●	**Fluid Conversion**
Fluid Model	OIL2 ●	Ignore GOR ●
Separator Data	Stock Tank Separator ●	Ignore WCT ●
Temperature	*GOR*	*WCT*
DegF	*scf/STB*	*Percent*
150.00	400.00	3.00
Pressure Control		
Constraint	*Lower Bound*	*Upper Bound*
	psia	*psia*
Pressure Bound ●	300.00	310.00
Fluid Control		
Constraint	*Lower Bound*	*Upper Bound*
	STB/Day	*STB/Day*
Oil ●	1000	2000
OK	***Cancel***	***Help***

Fig. 3–31. Source object fluid description

The purpose of the *tuning buffer* is to help solve pressure and/or temperature mismatch problems in un-tuned models. Once the model is tuned, and the mismatches eliminated, the tuning buffers would be switched off.

The *gas purifier* object allows a single-library inorganic component to be removed from a single-phase gas stream. This is a directional object.

The object uses the same simplified removal mechanism for each inorganic species, which is not dependent upon geometry or any "service" species. The object is both isobaric and isothermal (i.e., $P_{inlet} = P_{outlet}$; $T_{inlet} = T_{outlet}$).

There are four possible settings for gas inorganic removal: all inorganics (default), N_2, H_2S, and CO_2.The object is available for both black-oil and compositional fluid models.

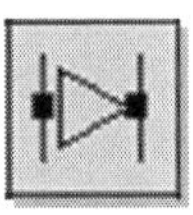

The *flow direction* object works like a non-return valve. This allows users to impose a direction of flow in a section of network. There is no pressure or temperature change across the object, and the arrowhead indicates the intended flow direction. The object itself has no editable parameters. The object will allow any flow rate (including zero) to pass through in the indicated direction, but will always try to prevent any flow in the reverse direction.

Other special equipment such as a multi-phase booster pump, jet pump, and eductor may also be required in the network model, based on specific field requirements.

3.5 Detailed Compressor Modeling

A description of detailed compressor modeling and calibration is presented in this section.

In a simple compressor model, compressor behavior is thermodynamically modeled as a function of throughput, compression ratio, horsepower, and efficiency. The physical design of the equipment is not taken into account. Neither the speed nor the performance curve of the individual compressor stage is modeled in simple compressor models.

A *detailed compressor* object is used to model the real performance of centrifugal compressors powered by gas turbines. The detailed compressor modeling involves modeling of the gas turbine, gear box, and compressor stages.

- On the gas turbine side, it estimates the horsepower available, using the actual heat rate or efficiency surface of the gas turbine to decide the horsepower generated against the speed. It also determines the quantity of fuel gas required by the gas turbine.
- On the gas compression side, it uses the compressor stage performance curves and determines the head generated for the gas throughput based on the speed of the compressor, ensuring that the flow rate is between the "surge" and "stonewall" limits.

Even though the design horsepower of the unit is fixed, the performance of the compressor train changes due to age and other operating conditions. Using detailed compressors makes it possible to get a realistic estimate of the compression output available at any time, taking into account the real-life constraints and performance degradation of the compressor units.

The detailed compressor model can work for both the black-oil fluid model as well as the compositional fluid model used for process gas. Using fuel gas properties, the solution will also estimate the quantity of fuel gas used by the unit.

The detailed compressor model should be built and tuned separately. Once the stand-alone model is calibrated and matched to the field conditions, it can be incorporated into the network system model.

Detailed modeling can also be performed for centrifugal compressors driven by motors.

3.5.1 Compressor Performance Curves

The compressor performance maps recognize the head developed versus gas flow rate, and the surge and stonewall limits of each casing.

Surge is an unstable flow condition that occurs at low flow rates. The surge limit is characterized by a cyclical change in the direction of flow; gas surges backwards and forwards through the compressor. Operating the compressor in surge condition will destroy it.

The *stonewall* limit is reached when the total flow becomes too high and the compressor chokes. The compressor performance curves with the surge and stonewall limits are shown in **Fig. 3-32**. The compressor should be operated between these two limits.

The model determines the speed of the compressor that is needed to deliver a discharge pressure and flow rate required by the gas lift network. If the discharge flow rate (determined by the field demand) is less than the minimum flow rate required for the speed, it opens the anti-surge (recycle) valve and allows part of the gas to recirculate back to the suction of the stage. This is done to maintain the minimum

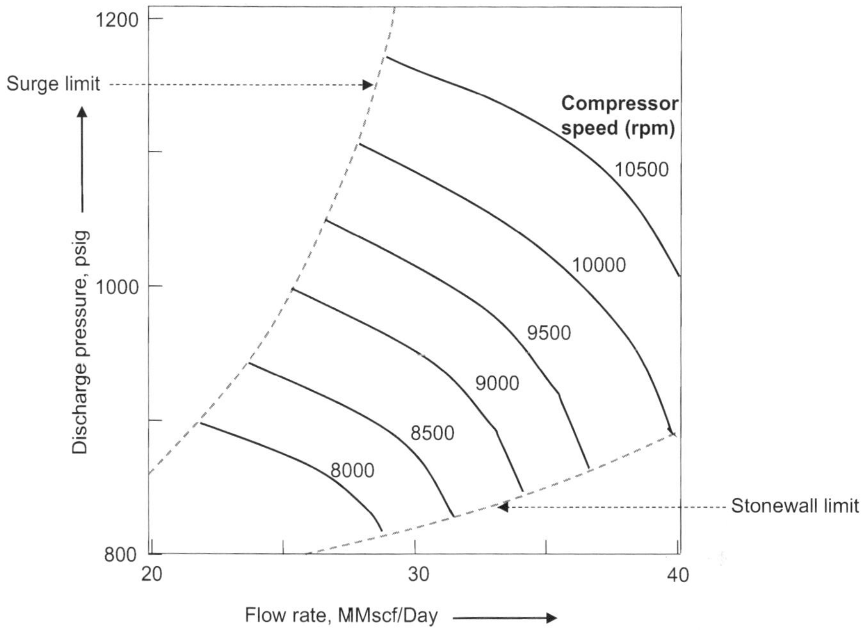

Fig. 3–32. Compressor surge and stonewall limits

required flow rate through the compressor barrel and to avoid operating the compressor at the surge point. However, this will reduce the efficiency of the system.

Another factor that can reduce the available horsepower is high ambient temperature, which reduces the efficiency of the gas turbine. At high ambient temperatures, the available horsepower decreases. The opposite happens during winter months. Companies running giant gas turbines devise production strategies to make the best of both these extremes between summer and winter, considering the horsepower variation. Detailed modeling of compressors has helped manage this very well.

In an asset that depends heavily on gas lift for its production, compressor horsepower and performance play a crucial role in optimizing production from the field. In mature fields with gas compression shortage, the field loses production whenever a compressor is shut down for maintenance. Thus, it is necessary to integrate the compressor maintenance program with the well management program to identify synergies. The detailed compressor model has offered immense benefit in this regard.

3.5.2 Compressor Output Determined by Network

A centrifugal compressor offers a reasonable level of flexibility to operate close to the design parameters of the machine. If the gas gravity is close to design, the horsepower required by the compressor will be decided by two variables: the compression ratio and the gas throughput. If the suction pressure is constant, the variation of the compression ratio means variation of discharge pressure.

For every discharge pressure, the compressor requires a minimum speed.

The actual output from a gas lift compressor (discharge pressure and gas flow rate) is determined by the gas lift requirement of the network, obviously limited by the maximum horsepower available from the prime mover.

Fig. 3-33 depicts the performance curve of a centrifugal compressor. At any given speed, it can operate with a range of gas throughput, within the surge and stonewall limits. Higher discharge pressures require higher speeds of the machine. At high discharge pressures, higher flow rates are required to avoid getting closer to surge limits for that speed.

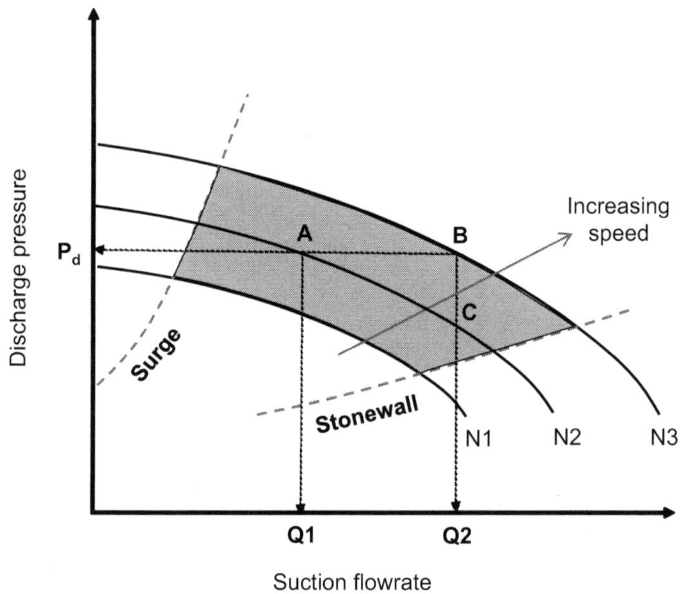

Fig. 3–33. Flexibility of centrifugal compressor

It is clear that a discharge pressure of P_d cannot be achieved with a compressor speed of N1. It is necessary to increase the speed to N2. With a speed of N2 and a discharge flow rate of Q1, the compressor operates at a safe level (point A).

If a flow rate less than Q1 is required, the machine speed should be decreased. Decreasing the speed will make the compressor operate on a different performance curve. If the machine is not slowed down, the operating point will move closer to the surge line.

If a higher flow rate (Q2) is required at the same discharge pressure (P_d), the compressor speed should be increased to N3 (point B). If the speed is maintained at

N2, a higher flow rate can be obtained, but at a lower discharge pressure (point C). Drawing too much gas flow without increasing the speed moves the compressor closer to its stonewall limit.

As the gas lift compressor and the gas lifted well network are incorporated in a single network model, the desired solution will be the one that maximizes oil production with the use of optimum compressor horsepower. Also it should allow the safe operation of the compressor for many years. The limitations and the range of safe operating parameters of the existing compressor should be clearly understood when performing optimization of gas lifted fields.

3.5.3 Modeling of the Gas Turbine

A gas turbine, also called a combustion turbine, is a type of continuous and internal combustion engine. The two main elements common to the gas turbine engine are: (1) the gas generator (GG), which includes the upstream air compressor, combustor, and a downstream turbine on the same shaft, and (2) a power turbine (PT) connected to the process compressor through another shaft.

Fig. 3-34 shows the schematic of a simple-cycle, two-shaft gas turbine unit. In this type of unit, the power turbine rotor is mechanically separate from the high-pressure turbine rotor. This feature allows the power turbine to be operated at a range of speeds and makes two-shaft gas turbines ideally suited for variable speed applications. [19]

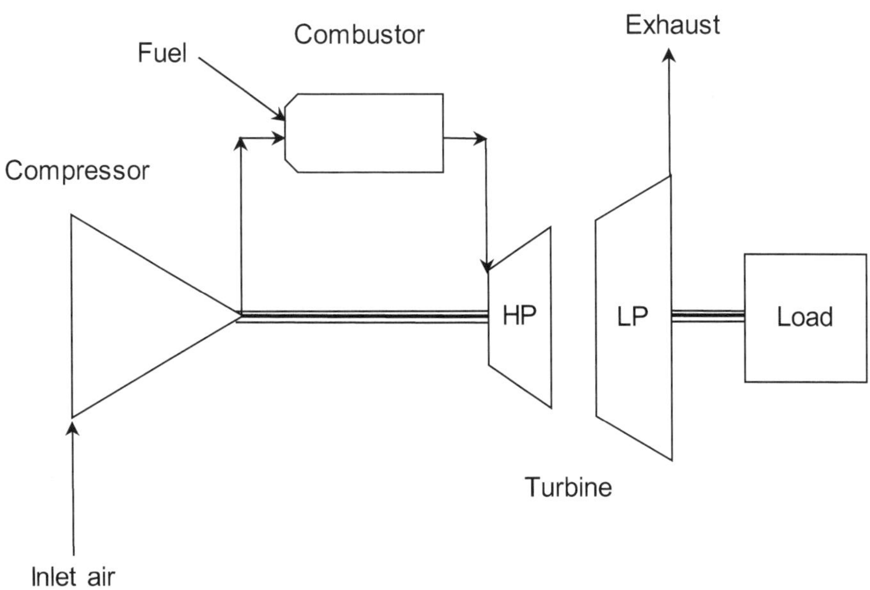

Fig. 3–34. Gas turbine unit

3.5.1.1 Gas Turbine Components

Gas generator: The gas generator consists of an axial compressor, combustor, and a high-pressure turbine connected by a rotor. The compressor of the GG section draws in air through the inlet plenum and compresses it. Fuel is mixed with compressed air in the combustor, where it burns and releases hot waste gases. The high pressure waste gases are expanded through the high-pressure turbine, causing it to turn. The GG is rotated at speed N1 by burning the inlet gas as fuel.

Power turbine: The unused energy coming out in the hot exhaust gases from the GG rotates a second independent turbine, called a power turbine, at speed N2. The PT is connected to the process compressor (load) via a gear box. The PT provides the power required for compressing process gas. The PT is not mechanically coupled to the high-pressure turbine.

Usually, one stage or two stages (LP and HP) or three stages (LP, MP, and HP) of compressor casings are connected to the gear box. Each casing may have several compressor wheels (stages), depending on the compression requirement. Between the stages, the gas will be cooled, and liquid will be knocked out of the gas stream before it enters into the compressor again. The whole unit, i.e., gas turbine, gear box, and the multistage compressor housings, is called a *compressor train*.

Fig. 3-35 shows the detailed model of a two-stage compressor. Note that each compressor stage is provided with its own recycle valve for anti-surge protection.

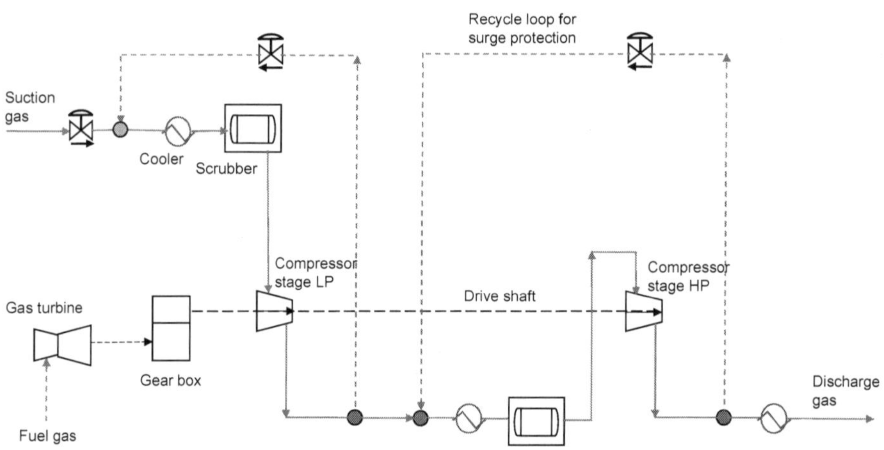

Fig. 3–35. Two-stage detailed compressor schematic

3.5.1.2 Gas Turbine Performance

The horsepower available from the gas turbine is calculated as follows:

The gas turbine (GT) object represents the combined performance of the gas generator and the power turbine. The curve that defines the upper edge of the gas turbine operating range gives the maximum shaft horsepower at the gearbox (GB SHP) available for a power turbine speed N2. The model should be able to predict the maximum power available on any day using ambient temperature. The increasing heat rate shown in **Fig. 3-36** is the same as decreasing efficiency. This means that the efficiency of the engine is maximum at the design speed.

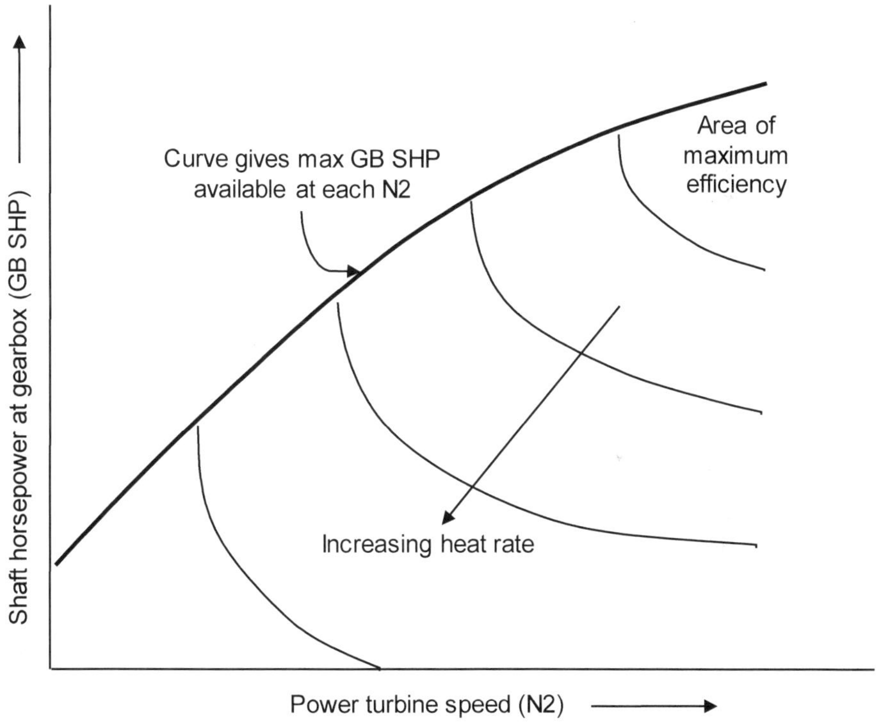

Fig. 3–36. Gas turbine efficiency

For a particular GT speed, there is a maximum power output, governed by several factors, including the maximum exhaust gas temperature and the maximum speed of the gas generator. The GT performance is represented as a function of power output and power turbine speed.

The turbine efficiency is the reciprocal of the specific fuel consumption of the GT (also equivalent to the heat rate), which is supplied by the manufacturer as a performance map.

The efficiency values for a GT can also be entered in the form of heat rates. As the speed decreases, the heat rate (or the specific fuel consumption) increases and thus the efficiency decreases (**Fig. 3-37**).

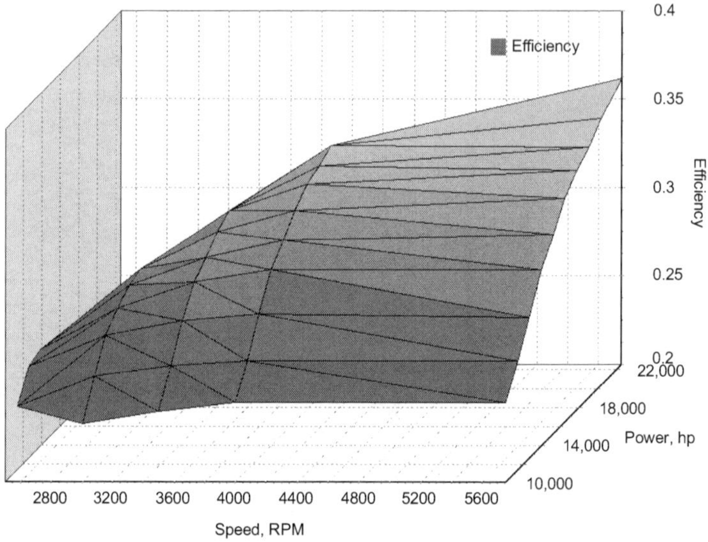

Fig. 3–37. GT Performance model

3.5.1.3 Estimating Available Horsepower

1) Estimation of PSHOP

The power generated by the gas generator (GG) is expressed as a relationship between the power measured in the shop (PSHOP) and the equivalent ISO GG speed ($N1_{ISO}$). The polynomial coefficients describing the power generated by the GG are available from the shop tests performed before turbine delivery, following manufacture or overhaul. Using these coefficients, the model will estimate PSHOP.

2) Calculation of PSITE

The model will apply corrections to PSHOP for the throat area and ambient temperature to obtain PSITE, the actual power delivered by the GG in the field.

The horsepower available at the site from a GG, calculated for an ambient temperature range of 60°F to 120°F was plotted against temperature. The plot is almost linear, showing a decrease of about 11% of horsepower for an increase of ambient temperature by 25° on the Fahrenheit scale (**Fig. 3-38**). This is in agreement with the results observed in the field. On a very hot day, the gas turbine output power is the least.

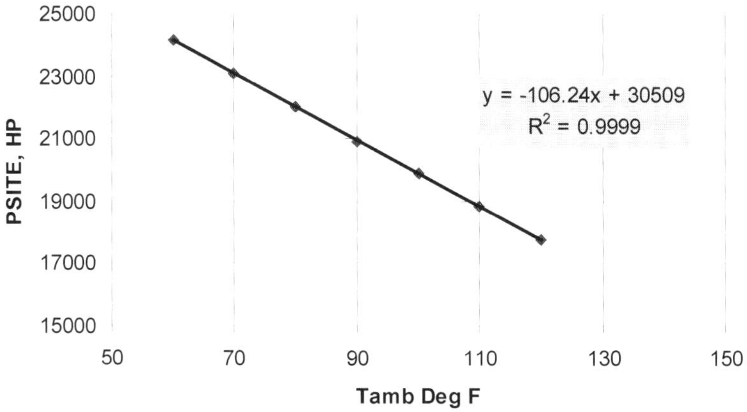

Fig. 3–38. Impact of ambient temperature on gas turbine output power

3) Calculation of PT SHP

From the actual power delivered by the GG in the field (PSITE), the power turbine shaft horsepower (PT SHP) is calculated by multiplying the PSITE by the power turbine efficiency (PT EFF), and then applying correction factors for atmospheric pressure (CFP_{ATM}) and the discharge pressure of the air compressor in the GG (CF_{CDP}). These correction factors are available from the GT performance test. This gives the PT SHP that is available at the gear box.

$$PT\ SHP = PSITE \times PT\ EFF \times CF_{PATM} \times CF_{CDP}$$

4) Calculation of GB SHP

The *gear box* is described by the gear ratio (ratio of the output and input speeds of the gearbox) and its efficiency (say, 98%). The shaft horsepower at the gearbox (GB SHP) is calculated using the gear box efficiency.

$$GB\ SHP = PT\ SHP\ x\ 0.98$$

This is the power available for the actual compressor stages.

3.5.4 Modeling of the Compressor Stages

A compressor stage is defined as a compression unit that may represent either a single impeller or, more often, a group of impellers in a single casing, modeled by the polytropic model. Typical values for polytropic efficiency (η_p) are given below:

Centrifugal compressor: 0.72–0.80

Reciprocating compressor: 0.92–1.0

To ensure that the model is accurate over a wide range of operating conditions, the performance curves are expressed as cubic polynomials.

Compressors are described using two performance curves: polytropic head coefficient, μ_p, and polytropic efficiency, η_p. These are plotted against the flow coefficient, ϕ.

These values are entered to create the compressor model. Each compressor casing will have its own performance curve, as shown in **Fig. 3-39**.

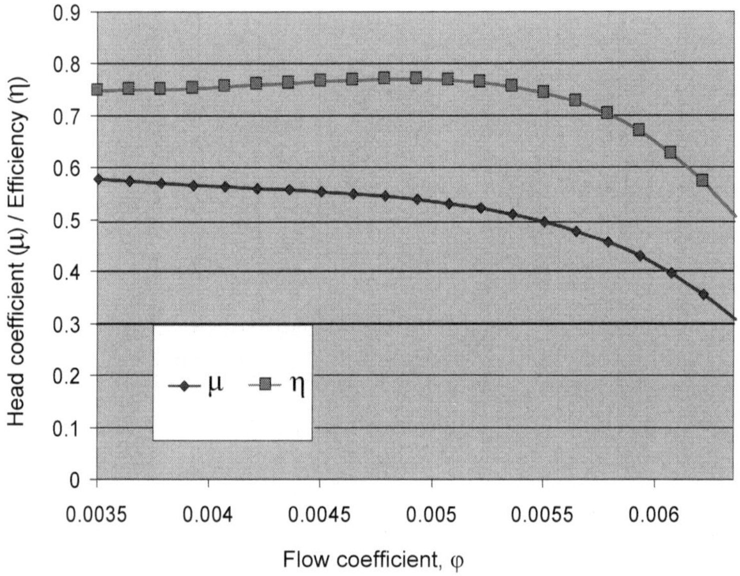

Fig. 3–39. Compressor model performance

3.5.5 Combining Gas Turbine and Compressor Performance

The performance curves of a centrifugal compressor can be plotted as head developed versus flow rate. Head developed is plotted on the vertical axis against flow rate, plotted on the horizontal axis. When plotted against suction flow rate, for every speed (N), there is a head versus flow rate curve for the compressor, the minimum flow rate being the surge limit. The compressor performance curves can be plotted for different speeds N_1 to N_5, noting that $N_5 > N_4 > N_3 > N_2 > N_1$.

The horsepower requirement of the compressor will also increase with speed. The horsepower can be plotted on the vertical axis versus flow rate on the horizontal axis. This graph shows the horsepower required by the compressor for various speeds N_1 to N_5.

With the common horizontal axis, **Fig. 3-40** combines the compressor performance with its horsepower requirement at various speeds.

At a speed N_4, at a specific suction pressure, the compressor can compress a certain flow rate (Q_4) and deliver at a discharge pressure of P_{d4}. To perform this operation, the compressor requires a horsepower P_4 as shown in the figure. The horsepower so needed should be available from the gas turbine.

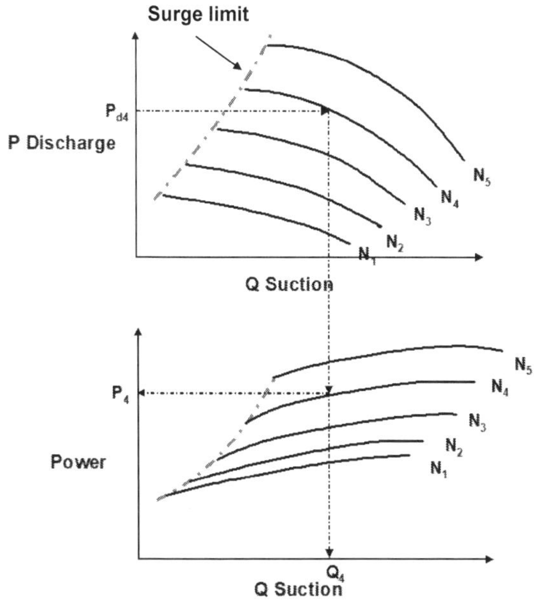

Fig. 3–40. Compressor head and horsepower requirement

The curve in **Fig. 3-41** represents the *maximum* shaft horsepower available at the gearbox (GB SHP) plotted against the power turbine speed (N2). The actual power used by the compressor unit would normally be in the area below this curve (area marked as *operating region*). If the GG is undersized or underperforming, the operating conditions of the GG will limit the horsepower generated, and this should be taken into account in the performance map. This is achieved by applying a cutoff to the power available, based on the constraints of the GG in service, such as max GG speed and high exhaust gas temperature. This is indicated by the horizontal dashed line *GG SHP @ GG Limit* in the figure. The horsepower curve beyond the cutoff point should be considered invalid.

When the *minimum horsepower* values (corresponding to the surge points) are plotted against compressor speed, a curve as shown in **Fig. 3-42** is obtained. It is seen that, at the higher end of the speed values, the minimum power curve is almost a straight line, power increasing with speed. For safe operation of the unit, the compressor is operated away from the surge line, and therefore the actual power required will be above that represented by the minimum power line. This area is marked as the *operating region* in the plot.

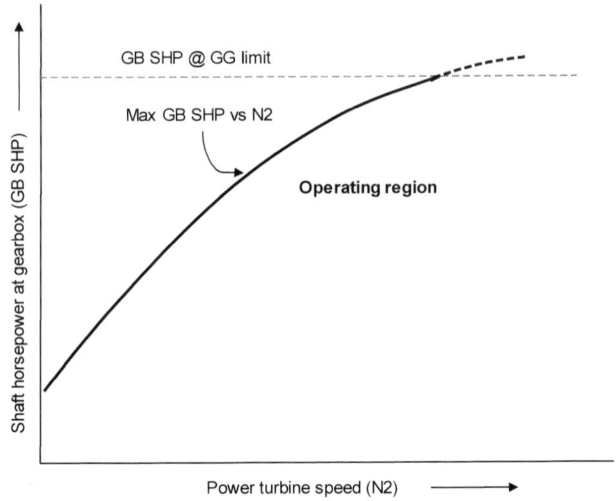

Fig. 3–41. Maximum horsepower available from gas turbine

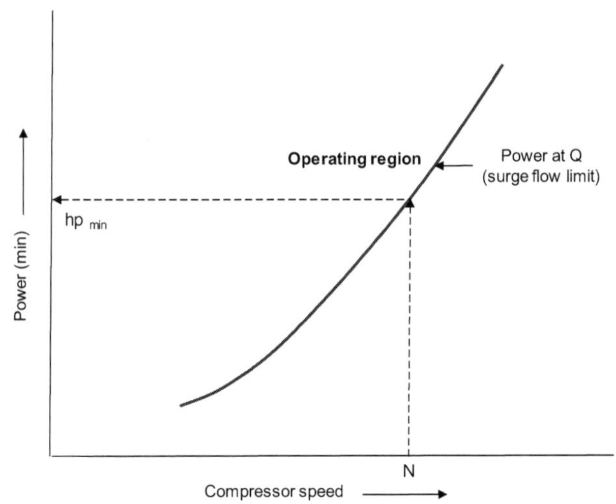

Fig. 3–42. Minimum power vs. compressor speed

The power turbine and the compressor are connected by a gear box and have a fixed speed ratio. Therefore, it is possible to combine these two figures: **Fig. 3-41** (power available from the gas turbine unit) and **Fig. 3-42** (power required by the compressor). The combined compressor and gas turbine performance is shown in **Fig. 3-43**. It is clear that the operating region falls in between the two curves—the maximum power vs. PT speed curve of the GT and the minimum power vs. speed curve of the compressor.

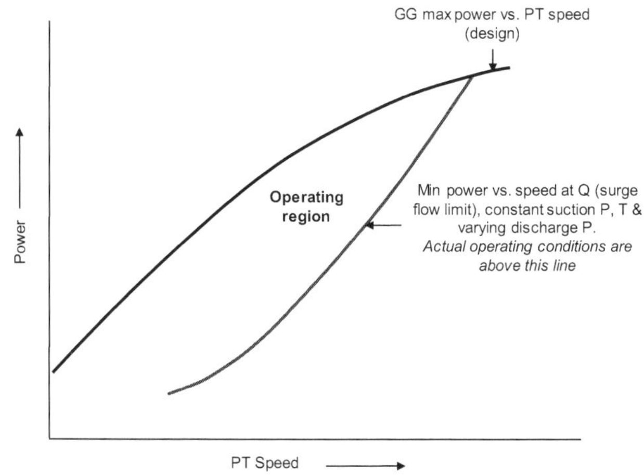

Fig. 3–43. Combined gas turbine and compressor performance

In gas lift optimization situations, considering the well performance and the size of the production and gas lift injection pipeline networks, a minimum gas lift header pressure is required to operate the gas lift network to achieve optimum production. This minimum gas lift header pressure controls the minimum compressor discharge pressure. As already concluded, there is a minimum compressor speed required for achieving this minimum discharge pressure. This speed can be shown in the combined GT and compressor performance diagram, indicated as a vertical line in **Fig. 3-44**. Note that, after imposing the minimum gas lift header pressure requirement, the operating region becomes smaller than before.

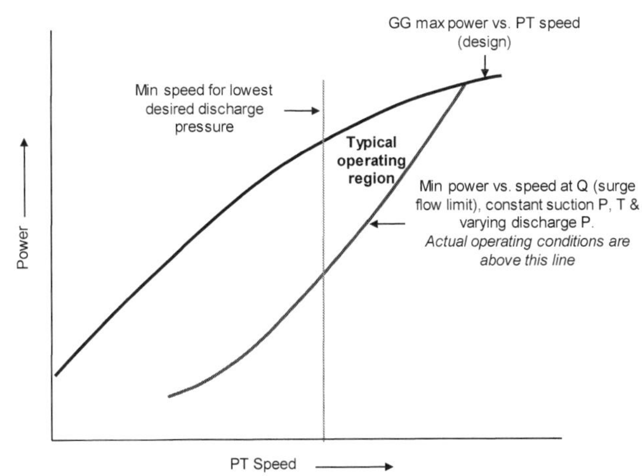

Fig. 3–44. Combined unit performance with gas lift pressure constraint

This minimum gas compressor pressure corresponds to a gas lift header pressure assumed in the field. All well gas lift designs are based on this pressure.

3.5.6 Example Modeling of Detailed Compressor

In this section, a detailed compressor modeling, performed for a pipeline compressor in an offshore field, is illustrated. This is a single-casing compressor.

The set of heat rate curves provided by the gas turbine manufacturer (**Fig. 3-45**) was digitized to obtain heat rates for various speeds N2 versus horsepower generated.

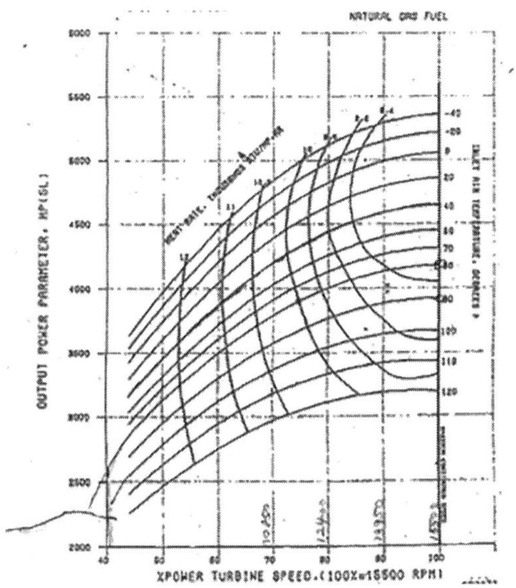

Fig. 3–45. Heat rate curve

The values from the digitized gas turbine performance surface were input into the software to model the GT. This defined the horsepower that can be obtained from the GT, plotted against the PT speed (N2), as shown in **Table 3-4**.

Table 3–4. GT performance table

	Speed	7502	8401	9436	10633	11470	11842	12400	12741	13516	15500
Power											
3250		12500	11900	11100	10575	10300	10250	10100	10025	9920	9800
3500		-1	11850	11000	10450	10200	10100	9960	9880	9780	9670
3750		-1	-1	10938	10400	10100	9960	9840	9780	9650	9520
4000		-1	-1	-1	10350	10000	9890	9750	9630	9550	9420
4125		-1	-1	-1	-1	9950	9840	9700	9610	9490	9333
4188		-1	-1	-1	-1	-1	9825	9680	9580	9440	9300
4250		-1	-1	-1	-1	-1	-1	9650	9550	9400	9250
4313		-1	-1	-1	-1	-1	-1	-1	9530	9380	9170
4375		-1	-1	-1	-1	-1	-1	-1	-1	9350	9100
4456		-1	-1	-1	-1	-1	-1	-1	-1	9000	9000

The temperature correction factor was calculated using the heat rate curves (**Fig. 3-46**).

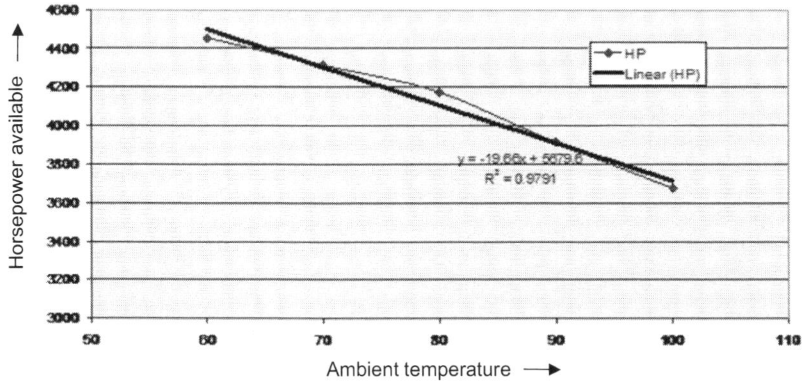

Fig. 3–46. Ambient temperature correction

The design compressor performance curves were digitized and used in constructing the compressor model. The coefficients for pressure and efficiency were used to describe the compressor performance (**Fig. 3-47**).

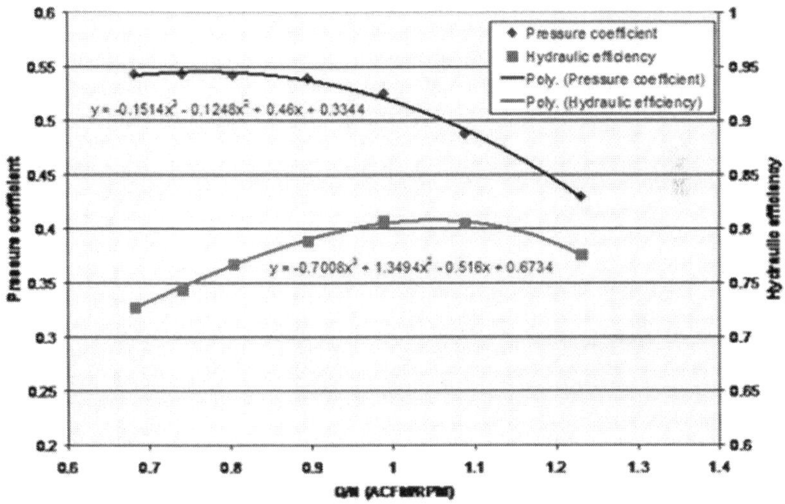

Fig. 3–47. Digitized compressor performance curve

After building the model, small adjustments were made to the head and mechanical coefficients so that the head developed and the mechanical efficiency of the compressor model would accurately match with the observed performance of the real-world compressor (**Fig. 3-48**).

Mechanical Efficiency			**Tuning Factors**	
Mechanical Efficiency	0.78	Fraction	Head Coefficient Tuning Factor	0.943
Equivalent Diameter	47.0	In.	Efficiency Tuning Factor	1.00
			Mechanical Tuning Factor	1.085
			Flow Coefficient Bounds	
QND3 Form ●			Surge Limit	4.7513e-003
QN Form ○			Stonewall Limit	8.0765e-003
Min Speed	3000	rpm		
Max Speed	7000	rpm		
Min power	2000	hp		
Max power	6000	hp		
Polynomial Coefficients				
Head			**Efficiency**	
Constant	3.344e-001		Constant	6.734e-001
Linear	6.8009e+001		Linear	-7.6288e+001
Quadratic	-2.7279e+003		Quadratic	2.9496e+004
Cubic	-4.8927e+005		Cubic	-2.2647e+006
OK		Close		Help

Fig. 3–48. Compressor stage modeling

The compressor model schematic is shown below (**Fig. 3-49**):

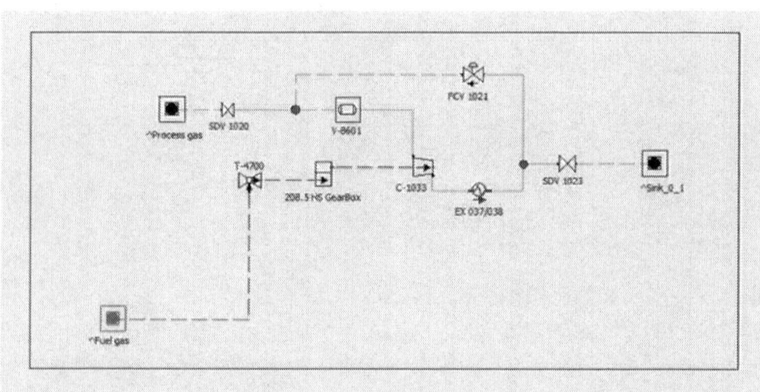

Fig. 3–49. Single-stage detailed compressor

This completes the building and calibration of the detailed compressor model.

The full set of results from the model after calibration is presented below (**Table 3-5**). Note that under current conditions the recycle valve is kept open by the optimizer to avoid surging, as the gas take off is less than the minimum flow required across the compressor. In other words, the compressor speed should be reduced to close the anti-surge valve. The discharge pressure will decrease slightly.

Table 3–5. Calibrated compressor model results

	Description	Compressor model	Field data (Used for calibration)
Gas turbine	Max power at ambient temp, hp	3729	
	Result power at ambient temp, hp	3729	
	Power %	100.0	
	Max speed, RPM	15500	
	Result speed RPM	14720	14730
	Speed %	95.3	
	Fuel gas rate, MMscf/Day	0.480	
Gear box 1	Ratio	0.414	
	Mechanical efficiency, %	1.0	
	Input power, hp	3729	
	Output power, hp	3729	
	Input speed, RPM	14720	
	Output speed, RPM	6095	6095
Compressor stage	Result power, hp	3729	
	Result speed, RPM	6095	6095
	Surge limit	0.00475	
	Result flow coefficient	0.00650	
	Stonewall limit	0.00808	
	Surge/Stonewall %	40.8	
	Suction pressure, psia	145	145
	Discharge pressure, psia	270	270
	Suction temperature, °F	135.7	139
	Discharge temperature, °F	223.6	227
	Flow rate, MMscf/Day	81.7	81.6
	Recycle flow rate, MMscf/Day	5.0	5.0

	Description	Compressor model	Field data (Used for calibration)
Package output	Suction flow rate, MMscf/Day	76.7	76.6
	Suction pressure, psia	145.3	145
	Suction temperature, °F	139	139
	Discharge press exit package, psia	265.1	268
	Discharge temperature °F	156.6	165
	Discharge flow rate, MMscf/Day	76.7	76.6

4

Running the System Optimization Model

Once the network model is ready, it is necessary to explore innovative ways of using the model. The benefits are obtained from day-to-day optimization evaluation and project investigation. People in charge of running and maintaining the model should have thorough knowledge of the production and process operations of the asset. They should be trained in the essentials of the modeling technology so that they can perform simulations to history match the model, set up optimization runs, and analyze results. In-depth knowledge of the model goes a long way in gaining extra value from it.

This chapter presents a discussion of the following topics:

- Preparing for running the model
- Setting up different types of objectives and constraints
- Performing history match
- Running optimization investigations
- Establishing system optimization process workflow

4.1 Preparations for Model Run

4.1.1 Bulk Data Entry and Editing Tools

The model being populated with a large number of objects, there is a great deal of data entry, some of it repetitive in nature. Accessing the screen of each piece of equipment and editing item by item is almost impossible for most of the models. The software usually allows the user to enter or edit the data in a bulk mode to save time.

Most of the models allow a table view for data entry or editing information. Table view data entry may be selected using either *sheet table view* or *global table view*. Sheet table view lists one type of equipment (e.g., pumps) available in that sub-sheet only, while global table view provides a list of the same from the entire model.

As an example, if it is necessary to implement a change in the cost of power for all the compressors in a model, the data of all the compressors can be displayed using global table view, and the value in the cost column can be edited. This value can be copied to all cells above/below it using the *Fill Up/Fill Down* commands, respectively.

Similarly, the metallurgy or insulation thickness of all the pipes can be changed by a single bulk data entry.

4.1.2 Assigning Revenue and Costs

As the optimization model works with the objective of maximizing net revenue ($/sec) for the entire field, the revenue and costs should be assigned for every producer and injector well, source, and sink, and every equipment that will need power. This will enable the optimizer to consider total revenue and costs of running the field as it optimizes the network.

In oil producers, it is normal to use prices for oil and gas produced and costs for water produced. Example values are shown below:

Gas: Price $500/MMscf

Oil: Price $30/BBL

Water: Cost $0.50/BBL (or Price $-**0.5/BBL**)

In well objects, the price of oil produced should be specified along with the cost of water produced. Even if the field has excess treating capacity for processing the produced water, it will cost money to treat and dispose of the water. If the oil is priced relatively low (say, $20/BBL), and with a high disposal cost for water (say $1/BBL), the model may selectively minimize fluid production from high water cut wells. This way, the field water production will be restricted without seriously affecting oil production. Most of the oil production will come from low water cut wells.

When using the source object for lift gas or fuel gas to be imported from another field, if the gas is free, assigning a very low cost value for the gas would be sufficient in the source object. But if the gas import is not free, then the real costs should be assigned so that the model will manage the import considering the spending.

4.1.3 Parallel Equipment in the Model

4.1.3.1 Wells Requiring Differentiation

If wells with different fluid compositions are connected to a network, and if it is necessary to curtail the production of one type of fluid, the wells should be differentiated in some manner so that the optimizer can understand it. In some cases, gas produced by wells can be differentiated based on calorific value.

Considering two gas wells, P and Q, with well Q having a higher concentration of H_2S, one option is to use a lower price for the gas produced by well Q. This would encourage the optimizer to produce more gas from well P, based on the price difference specified, and at the same time, produce sufficiently from well Q to meet the export rates. Another option is to set an upper limit for the H_2S concentration in the produced network at the delivery point. The optimizer will restrict the production from well Q to honor the target constraint. The performance curves of the wells and limitations imposed by chokes and pipelines will always be considered during optimization.

4.1.3.2 Parallel Pumps or Compressors

There are situations where assigning some differential costs will help the model reach a solution quickly. A typical example is running parallel pumps or compressors. This should be considered only if optimizing the parallel equipment is needed.

When running parallel pumps, differentiating the cost of power will help the optimizer run the pump that requires lower cost. For example, if an installation has four identical main oil pumps, each with a capacity of 20,000 BPD, and if the field production is 50,000 BPD, three pumps are required to run when the field is producing at capacity.

In reality, if all the pumps are available for service, and the pumps and their piping are identical in performance, it does not matter which three pumps are running. In this case, it is of no benefit for the optimizer to spend time deciding among the pumps, as this choice does not add any value. Should this become important, introducing a small difference in the cost of power among the pumps will assist the optimizer to use the pump with lower operating costs and then move on with optimization. But if a pump is under maintenance in the field and turned off in the model, the optimizer will use the three pumps available for service, irrespective of the cost of power.

For such a situation, if individual pump modeling is not required, a *super pump*—a simple pump object with a capacity of 80,000 BPD—can be set up to pump the produced fluid. Even if the individual pumps have been modeled, the super pump can be connected using links on a bypass configuration and can be used when needed. When using the super pump, the flow through the four pumps should be turned off. In this case, the model will compute the cost of pumping and the horsepower required, but will not be concerned about which pumps are running or how many pumps are needed at any particular time. Note that the pressure drop in the suction / discharge lines of individual pumps will be ignored if a super pump is used.

The elements of optimization required should be the deciding factor for setting up a specific model configuration.

4.1.4 Phase vs. Total Fluid Optimization

With black oil fluids being used, either a total flow optimization or a phase optimization can be performed. If compositional fluid models are being used, only total flow optimization is possible.

Black Oil Fluids	Compositional Fluids
Total Flow Optimization OR Phase Optimization	Total Flow Optimization Only

In total flow optimization mode, the network will try to maximize total fluid production. Even in this mode, the optimizer will attempt to maximize oil production if the price of oil / gas and the cost of water are set at the wellhead object, but the underlying theme will be to increase the total fluid production.

If the fluid modeling has been done using the black oil description, the model understands the fluid movement as oil, water, and gas phases and inorganics moving in the network. So it would be preferable to use phase optimization to put emphasis on maximizing production of the oil or gas phase. The model also allows the user to place additional, different objectives/constraints on other phases. If an upper limit is set on the total amount of water produced, the phase optimizer will quickly honor the constraint set up on the water phase, even though the costs play a role in the process. As a result of this, the phase optimization may lead to faster convergence of a gas lift network model.

4.1.5 Model Validation

This is an operational check that the model performs to ensure its integrity. The model makes sure that there are no loose pipes anywhere in the system, and that all ends are connected properly. All wells and source objects should have fluid models assigned to them.

Every time an optimization run is initiated, the model will first perform a validation doing a check for the following:

- Missing connections
- Missing performance data
- Missing fluid information
- Inconsistent fluid data
- Active tuning objects

In case of such deficiencies, an error message will appear and the optimization will not proceed.

4.1.6 Snapshots

Optimization scenarios that can be investigated without changing the model construction are called *snapshots.*

Snapshots are copies of the physical network configuration, which allow different operating scenarios to be explored without changing the physical construction of the model. However, fluid data and equipment object attributes such as status (e.g., a block valve open / closed, a well flowing / shut-in) can be changed. Wells can be lined up to a different separator. Some of the input parameters (e.g., pipeline diameter, maximum diameter of choke) can differ among snapshots, but introducing a new pipeline will require a change in model construction.

Common examples of snapshots include:

- Run a snapshot with three of the four compressors running, and compare the results with the case of all four compressors working. This is to run a case for the maintenance of one compressor in the field.
- Line up some of the wells to a different production header (e.g., dry to wet header), and run an optimization investigation to understand its impact on overall field performance.
- Separate the commingled flowlines of two wells into individual flowlines (or vice versa), and evaluate the impact of pressure drop in the new configuration.
- Run a case with some lift gas import, and compare production performance of the field with no gas import.
- Run a scenario to find out the most economical way of increasing field oil production by 5% in a gas lifted network.
- Run a case to find out choke settings that will help increase production from the NF wells by 5%.
- Investigate the overall impact of upsizing a production trunk line from a satellite platform to the separation facility, and determine the optimum size of the pipeline required.

Snapshots can be edited and optimized, and their solutions can be reset independently of each's objectives and constraints. Optimizer parameters can also vary between snapshots.

From a single model with more than one snapshot, all the results can be extracted by giving a single command. The results from multiple snapshots can be compared between columns of a spreadsheet. This saves time when comparing the results.

4.2 Objectives and Constraints

4.2.1 Role of Objectives and Constraints

Setting proper objectives and constraints is a very important task. In the ideal world, it would be preferable not to have any constraints in the model and give full freedom to the optimizer. But it is necessary to honor the reservoir exploitation strategy, with the aim of improving the ultimate recovery from the reservoir. It is also necessary to factor in the real-life condition of an existing pipeline or piece of equipment, for maintenance or safety reasons. Objectives and constraints are extremely useful for this purpose.

Too many constraints set up in the model crimp the wings of the optimizer, and the model gets directed to move in a single direction. It is almost like running the model to get a pre-planned result. This will not unlock the full potential of the wells or the whole of the network.

The other extreme is complete relaxation that may make the optimizer wander around, not reaching any destination on time. The best approach would be to set up *minimum constraints*, as needed, in line with the optimization objective.

The users of the model should be thorough, with the right use of objectives and constraints. In general, the user should not set up non-realistic targets that the model cannot deliver. If the field cannot deliver it, the model also cannot do it.

It is mandatory that the user should, before setting up a new scenario investigation, examine the model and review the objectives and constraints already existing in the model. Some of these may be left in the model from a previous investigation. The user should review whether these constraints are still required for the present run and remove them if not required. This is a very common problem with large models. A large number of constraints (some of them may be conflicting) will impede the working of the model.

Some situations that will normally require constraints to be set up in the model are listed below:

- Honoring reservoir management strategy on well production (e.g., the maximum limit of production choke size, limiting the drawdown in wells, etc.)
- Setting up constraints required for history matching the model, to study pipeline or equipment performance, based on operating parameters received from the field
- Having real-world constraints of the field in the model (e.g., limiting the speed of a compressor due to vibration problems, limiting the working pressure of a pipeline due to integrity issues)

- Defining objectives and constraints that are required to convey specific goals of optimization to the optimizer. Examples:
  - Allow the model to find the best way to operate the field if the gas export should be increased from 20 to 40 MMscf/d;
  - Allow the model to determine the most efficient way to boost production by 5000 BOPD from the field.

It is possible that due to the large size and very complicated nature of some networks and the varied options available (e.g., parallel equipment, loops and branches, many recycling options), the optimizer may try to move in several directions and will be unable to get to a universal solution within the number of iterations permitted in a run. If a localized solution is given, the result may be sub-optimal in the context of the global opportunities of the full network and should be avoided. Minimizing the constraints may be beneficial in this case. The user should ensure that the constraints set are not functioning like a bridle for the model.

4.2.2 Types of Objectives and Constraints

In phase optimization with black oil fluids, the objective or constraint is usually applied for *pressure, flow rate,* and *gas concentration.* The constraint so applied could be one of three types:

- **Bound (Constraint)**:

  Allows the user to specify lower and upper bounds
- **Maximize/Minimize (Objective)**:

  Allows the user to specify lower and upper bounds and revenue and cost, respectively. If the item has revenue associated with it, the optimizer tries to maximize it; if it has cost, the optimizer tries to minimize it, within the overall context of the model.
- **Target (Constraint)**:

  Allows the user to specify lower and upper bounds, target value, target tolerance, and violation penalty

These are discussed in detail in the following sections.

4.2.3 Bound Constraints

The simplest of the three forms of constraint is the *bound* constraint. The user only needs to enter the lower and upper bounds of the allowable range of the variable. This type of constraint would ensure that the operating pressure of a separator is kept within a required limit (**Fig. 4-1**). The bounds are needed to convey the practical operating limits of the separator, as this affects the compressor suction pressure.

Link_8 ●	Link_10 ○	Link_11 ○	Link_13 ○
From: Separator Heat Exchanger		Reverse	Plot
To: Separator Stage 1			

Name	Value	
Pressure ●		
Type	Bound	
Apply To	Pressure	
Units	psia	
Lower Bound	200.00	psia
Upper Bound	220.00	psia
Flow rate ○		
Concentration ○		
OK	***Cancel***	***Help***

Fig. 4–1. Defining the bound constraints

It should be noted that, by setting the wrong bounds, the user might limit the real outcome of the model. This should be verified when reviewing the results to confirm whether any of the set bounds has influenced the decision of the optimizer. For example, in this case, if the model solution for separator pressure is 200 psia, it is possible that the optimizer may want to lower the separator pressure further. One more optimization run may be performed after resetting the bound constraints to 190 psia / 210 psia, and the results compared in terms of production and gas lift compression.

The logic behind bound constraints is shown in **Fig. 4-2**.

The bounds are considered “hard” limits. The penalty is zero between the upper and lower bounds, but rises steeply outside of the allowed range. In other words, the optimizer has the freedom to find a solution between these two limits. However, if a solution cannot be found between these two limits, the optimizer may settle for a solution outside, but will pay a huge internal penalty. This will be displayed in the results as a caution.

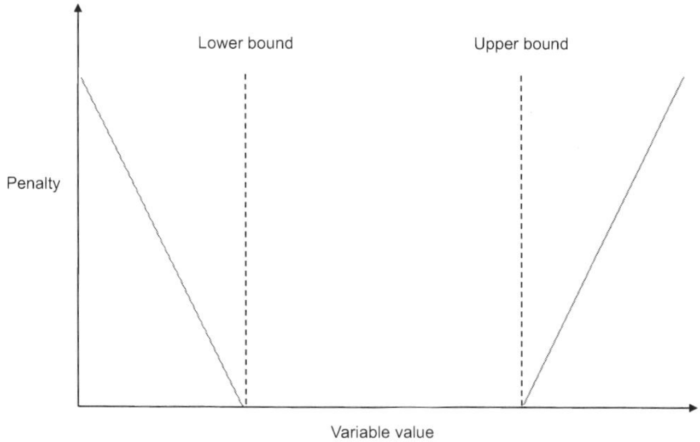

Fig. 4–2. Bound constraints logic

4.2.4 Maximize and Minimize Objectives

Maximize and *Minimize* objectives are normally used for entering revenues or costs associated with pressure or flow rate at that point in the network. For a maximize objective, the *revenue* (e.g., oil price) must be entered. For a minimize objective, *cost* (e.g., water disposal) must be entered, as shown in **Fig. 4-3**.

Link_15		
From: Choke_0	Reverse	Plot
To: Oil Storage		
Name	**Value**	
Pressure ○		
Flowrate ●		
Type	Maximize	
Apply To	Only Oil	
Units	STB/Day	
Lower Bound	0	STB/Day
Upper Bound	50000	STB/Day
Revenue	25.00	$/STB
Gas Concentration ○		
OK	***Cancel***	***Help***

Fig. 4–3. Setting maximize objective

In this case, the total reward or cost increases linearly between the lower and upper bounds (**Fig. 4-4**).

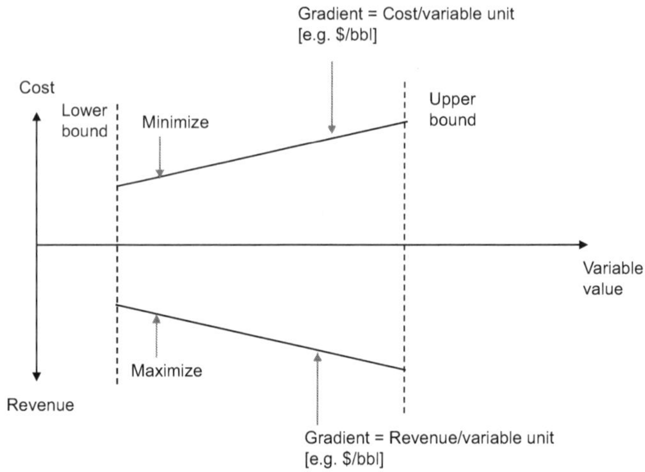

Fig. 4–4. Logic of maximize / minimize objective

4.2.5 Target with Penalty

The *Target* constraint allows the user to specify a target value or range for a variable, with a user-specified penalty for deviation within a second set of bounds. The user enters the lower and upper bounds, the target value, target tolerance, and penalty (**Fig. 4-5**).

This type of "target" is required to find the best compromise among conflicting objectives in a system. An example might be ensuring maximum production by driving down wellhead pressure in a gas field while maintaining optimum intake pressures to a compressor train.

In **Fig. 4-5** , the target is set to deliver at 6000 psia, with a target tolerance of 100 psi on either side. The outer boundaries, i.e., the lower and upper bounds, have been specified as 5000 and 7000 psia, respectively. In reality, this means that the result can be anything between 5900 and 6100 psia, but if it falls outside this tolerance limit, there will be a violation penalty of 0.01 $/(psi. sec).

How this works is shown in **Fig. 4-6**.

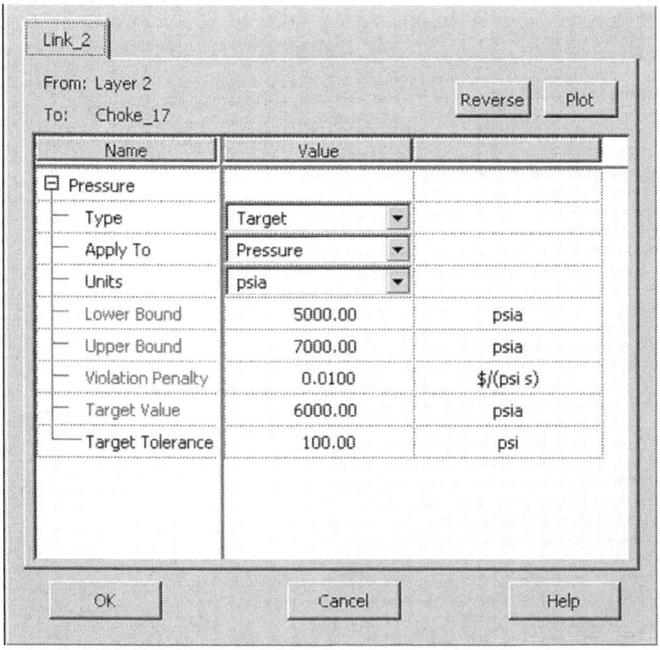

Fig. 4–5. Setting target constraints

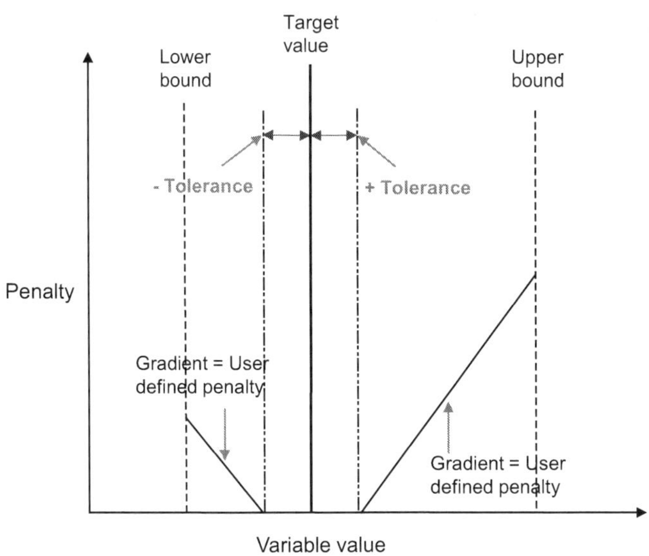

Fig. 4–6. Target constraint schematic

The target could be on flow rate, pressure, or composition. The user should assign what fluid type (oil, water, or gas) the target pertains to.

The controllers on the "Gas Concentration" tab affect gas quality by placing restrictions or requirements on the mole fractions of the following in the gas phase:

- All inorganics (CO_2, N_2, and H_2S)
- CO_2 specifically
- N_2 specifically
- H_2S specifically

Usually, the required effect is that the concentration of a particular component does not rise above a certain limit without incurring a penalty (**Fig. 4-7**). This effect is best achieved using bound or target constraints.

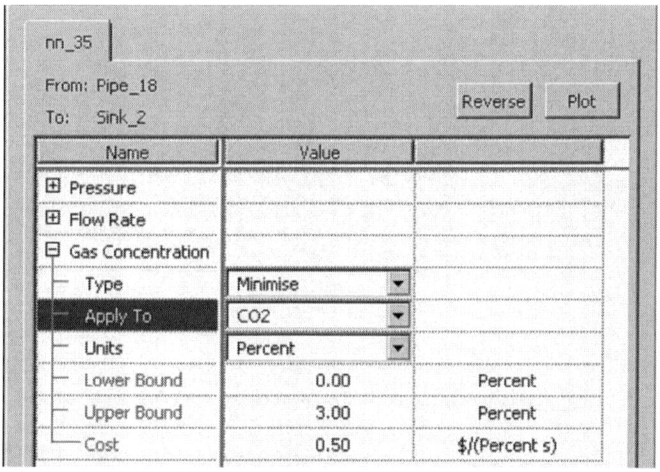

Fig. 4–7. Minimizing inorganic content

Concentrations are entered as either molar percentages or fractions of the gas phase at standard conditions. At standard conditions, all gases have the same molar volume, so volumetric terms are equally applicable.

4.2.6 Compositional Objectives

Similar constraints and objectives discussed above can also be set using total fluid basis.

4.2.7 Infeasible Solutions

When specifying deliverables from a model, the user should not ask for something that contradicts the physics of the flow or something outside the performance curve of the equipment.

Fig. 4-8 shows the performance curve of a centrifugal pump. If a higher head is required [marked (+) in the diagram], the pump can deliver it only if its speed is increased. At constant speed, fixing the *in-situ* flow rate fixes the head rise, and vice versa. If there is no mechanism to increase the speed, then the pump cannot deliver the head marked.

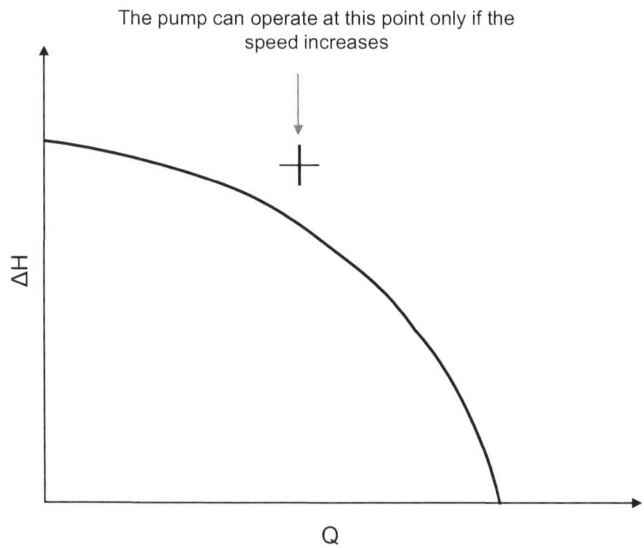

Fig. 4–8. Example of infeasible solution

The performance of centrifugal pumps at speeds other than the design speed is given by the Pump Affinity Laws (or Fan Laws):

$Q \alpha (N)$

$\Delta H \alpha (N)^2$

Power $\alpha (N)^3$

where N is pump speed.

The software should automatically re-scale the pump performance curves and establish the minimum and maximum flow rates applicable for different pump speeds.

Users should note that non-realistic objectives will not be honored by the model, or some limits will be exceeded to return with a solution.

Other examples of infeasible solutions are:

- Trying to optimize a gas lifted network after inadvertently closing the gas export valve
- Specifying a flow rate through a pipe or a choke such that, if delivered, the fluid velocity would exceed the velocity of sound. The model may not allow this to happen for safety reasons.

4.3 History Matching

Once the full network model is built, it should be calibrated to the current operating conditions in the field to replicate the performance of wells, pipelines, and equipment. This is called history matching.

History matching is the same as a constrained optimization. In preparation for history matching, the status of well, equipment, choke, and valves is set in the model as per field settings. Constraints and objectives are applied to replicate the field conditions. The model is allowed to calculate other parameters.

It is necessary to set all the constraints in the full model at once and run the history match.

4.3.1 Wells History Match

4.3.1.1 NF Wells History Match

A simple offshore production unit with naturally flowing wells will be considered. The wells are on a satellite platform producing into the production separator at the Facilities platform (**Fig. 4-9**).

The production from well B1 goes through its wing valve, flowline segment A, adjustable choke, flowline segment B, check valve, flowline segment C, production manifold D, then a long subsea pipeline E before it reaches the inlet header of the separator.

For the purpose of history match, accurate production rates are required from the wells. This can be done by calibrating all the well models with the latest well test data. Once the well models are calibrated, if the wellhead flowing pressures in the model are set as per field values, the production rates delivered from the wellhead objects, based on the well performance curves, will almost match with the field rates.

Once the model production rate is matched, the balance of history match boils down to checking the pressure drop in the pipelines.

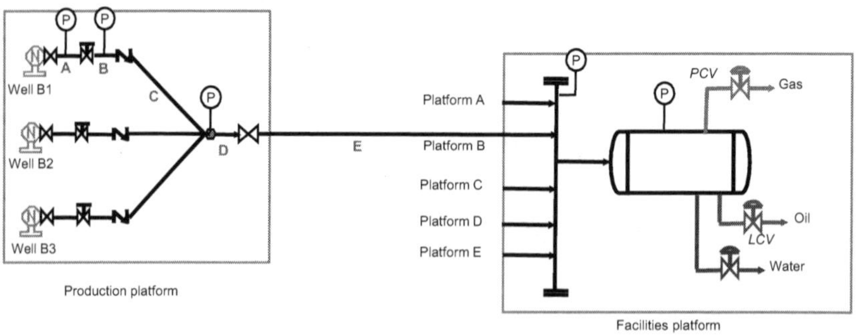

Fig. 4–9. History match for production system

If the model is run after fixing the wellhead pressures (thereby fixing the flow rates in this case) and the separator pressure, the model will calculate the pressure drop in the pipelines for the known flow rate.

The values calculated by the model are displayed in the highlighted columns in **Table 4-1**.

Table 4–1. History match results

	Pressure upstream of wellhead choke (FWHP), psig	Pressure downstream of wellhead choke, psig		Production manifold pressure at Platform B, psig		Production separator pressure, psig
	Field / Model	Field	Model	Field	Model	Field / Model
Well B1	800	275	260	265	255	125
Well B2	950	315	259	265	255	
Well B3	880	273	259	265	255	

The pressure at all nodes all along the pipeline in the model can be compared with field data if measurements are available. As the pressure at the delivery point (separator) is the same in the field and the model, the cumulative pressure drop in the direction opposite to the flow will be seen at the point downstream of the wellhead choke for every well in this case.

The pressure drop in the main departing pipeline is 140 psi in the field (130 psi in the model). Even if the pipeline is clean, a small difference in pressure drop is usually expected, as the pipeline may be old and the actual roughness may be different from what has been input into the model.

But there is something unusual in well B2, as the pressure drop in flowline sections B and C and the check valve is much more than calculated by the model. This could be due to a broken flapper of the check valve or some form of solid buildup in the flowline section. This would have been obvious even before the modeling was taken up and should be resolved in the field.

The chokes in all three wells are on critical flow. Correcting the well problem in well B2 may not lead to a significant increase in production at this time, but it will remove the additional restriction to the flow.

4.3.1.2 GL Wells History Match

For gas lifted wells, apart from the wellhead flowing pressures, the gas lift injection rates are also fixed using bound constraints in the model. It is a practice to give a very small window between the upper and lower bound constraints. If the lift gas rate is 2.5 MMscf/Day, the upper and lower bounds may be set at 2.60 and 2.40 MMscf/Day, respectively. With the well models already calibrated with the most recent well test data, well production rates from the model would almost match with the field rates.

The model will calculate the pressure and temperature throughout the network. The results are analyzed to examine how different the calculated pressures and temperature in the network are, compared to actual field data. The pressure drop across wellhead chokes, well flowlines, and production pipelines should be compared.

The sum of well gas lift injection rates fixed for history match defines the gas flow through the compressors, if all the compressed gas is used by gas lifted wells. The discharge pressure of the compressor and the pressure drop in the gas pipelines can be compared. The horsepower consumed in the field can also be compared with the value predicted by the model.

The wells, pipelines, and equipment items with significant mismatch in pressure drop should be investigated thoroughly. It is also possible that incorrect input data can cause a mismatch.

The history match would highlight if any of the pipelines is severely overloaded or badly plugged up. It will also identify pieces of equipment that are underperforming.

4.3.2 Equipment History Match

The history match of a simple water injection centrifugal pump of the produced water disposal system will be reviewed (**Fig. 4-10**). Two pumps are available. Pump A is in service, and pump B on standby.

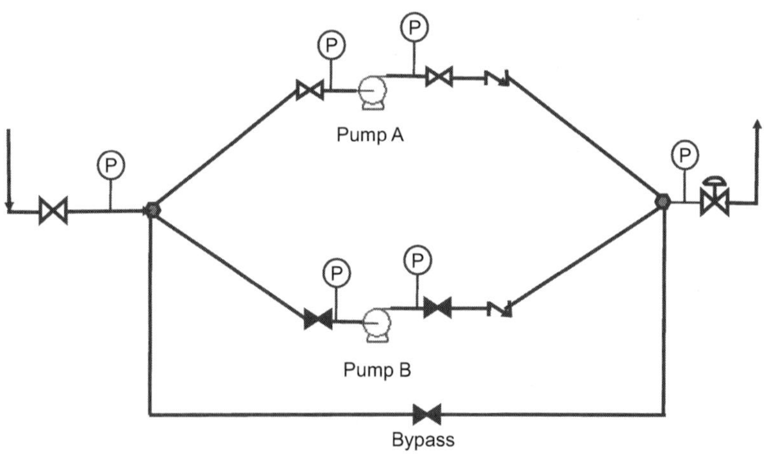

Fig. 4–10. Water pump in water disposal system

The prime mover of the pump is a 1000-hp electric motor running at constant speed. The water flow rate is 9000 BWPD. If the water flow rate is 9000 BWPD, as per pump performance curve, let us assume it should generate a head of 6900 ft. with a pump efficiency of 65%.

Under the current operating conditions, with a water specific gravity of 1.03, the pump should be adding 6900 ft. of head or 3077 psi, but, as per field data, it is adding only 2850 psi (6390 ft.). The head developed by the pump has deteriorated by about 7.4%. This will impact the volume of water injected by the disposal wells.

The results are tabulated in **Table 4-2.**

Table 4-2. Water disposal pump performance

	Pump rate, BWPD	Suction pressure, Field/ Model psig	Pump discharge pressure Field psig	Pump discharge pressure Model psig	Horsepower consumed in Field (HP)	Horsepower required in Model (HP)
Pump A	9000	100	2950	3177	775	671.2
Pump B	NA					

The water horsepower required is given by:

$$Water\ horsepower = \frac{q * h * (SG)}{3960 * \eta}$$

Where q = flow rate in gal/min

h = head developed in ft.

SG = specific gravity of water (fresh water = 1.0), and

η = pump efficiency (fraction)

For the head developed now (6390 ft.), with SG =1.03, the hydraulic horsepower required with the pump design efficiency of 0.65 would be 671.2 hp; but it is consuming 775 hp, as per field data. This confirms two problems: (a) deterioration of head developed and (b) big drop in efficiency, both requiring corrective action.

The excess power consumed is equivalent to 103.8 hp = 77.4 kW.

The excess energy consumption per day = 77.4 kW x 24 hours = 1857.6 Units of power. If the power cost is $0.5/kW-Hr, the excess bill is $928.8/day or $27,864/month.

4.3.3 Acceptable History Match Results

Upon convergence of the model, the user reviews the full results from the model for wells, pipelines, and equipment. Two options are usually available: summary results or detailed results. It is possible to extract important results into a spreadsheet so that the key results can be viewed without going back and forth in the model view. The model will compile the results category-wise (wells, pumps, compressors, pipelines, etc.) while giving output to Excel.

Another useful approach is a *rule-based visualization*. The user can set up simple rules in the model to highlight the following (a few examples are listed):

- Pipelines with mixture velocities more than, say, 50 ft/sec
- Pipelines carrying liquids that have velocities less than, say, 3 ft/sec
- Pumps using 100% of available horsepower
- Chokes that are in critical flow
- Chokes that went to a fully open position during the optimization
- Objects having results coinciding with upper or lower bound constraints

Normally, if the model fluid rates are obtained within +/-2% of the field values, the history match is considered good. A history match may be valid for several months if the field does not undergo major changes within the time period.

4.4 Optimization Scenarios

4.4.1 Instantaneous Optimization

Starting with a calibrated network model, an instantaneous network optimization can be performed at any time. This is the step that generates revenue for the company immediately.

The constraints applied for the history match should be removed. Specific optimization objectives should be introduced into the model for performing field optimization. After the optimization run is completed, detailed evaluation of the results is required.

Normally, during the project stage, the modeling consultants would conduct in-depth discussions with the production operations and engineering staff of the operating company and choose four or five optimization scenarios that are the most valuable for the operating company. This should add value for the instantaneous optimization in the field. After the project handover, the production engineering team would run optimization scenarios.

An optimization study of a simple gas lifted offshore field will be discussed briefly. The schematic of the field is shown in **Fig. 4-11**.

The following are the salient features of the field:

- The field has three types of wells—NF, GL, and ESP wells—distributed in various production platforms. The chokes on GL wells are fully open. The NF and ESP wells have adjustable chokes.
- The high-pressure gas from the compressors is mostly recirculated for gas lift injection into the wells. A small portion of the gas is exported.
- If the compressors are not running, the LP rich gas can be exported using a separate line so that the field can continue to produce. However, there is no facility to import lift gas.
- There is also a provision to import lean gas to be used as fuel gas for the compressors. If this is not available, inter-stage gas can be used as fuel for powering the gas turbines.

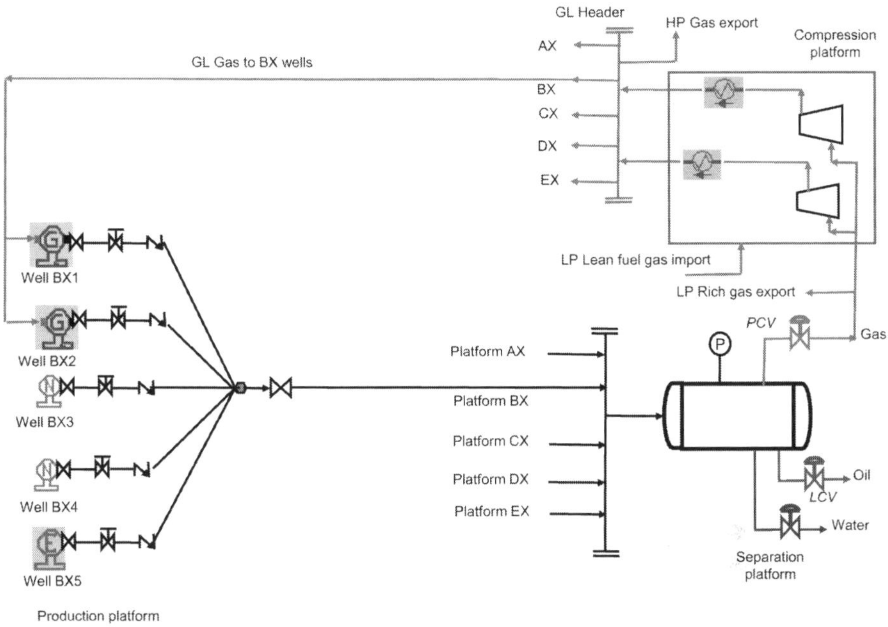

Fig. 4–11. Simple gas lifted offshore field

Some of the optimization scenarios that can be run for this field are indicated below. Several investigations would be required and are possible, based on the operating condition of the field and the operational requirements of the overall asset.

1) Optimize oil production and allocate gas lift to the wells in the field. The solution should come up with the best separator pressure, compressor discharge pressure, and well gas lift rates for normal operations; choke settings for the NF wells; and power frequency for the ESP wells.
2) Increase the high-pressure gas export by a certain volume (this is possible only if more gas enters into the network—preferably by producing more oil from NF and ESP wells or gas wells, if available).
3) Optimize production from the field when one gas compressor is taken off for engine overhaul maintenance (3–4 days downtime).
4) Maximize production from the field if ESP wells are not running for a week.
5) Optimize production from the field when the water production has to be limited to a certain volume (due to produced water treatment problems).
6) Ad-hoc operation to increase oil production from the field by 10% for 15 days to meet production target for a quarter.

The production optimization team would run such investigations, analyze the model results, confirm the results are practically valid, and document the findings as a *production strategy*, keeping it ready for implementation. This would be updated on a regular basis.

In an offline environment, small to medium-sized network models can be updated once a month. In large fields having several hundreds of wells, updating all the well models manually on a monthly basis would be a challenge. An automated model update would be needed, and this is best done by real-time optimization systems.

4.4.2 Gas Lift Optimization Solution by Optimizer

A gas lift response curve generated for a single flowing wellhead pressure is used for a single well gas lift optimization. In a network, the model should consider the impact of all wells connected to the same production header. The flowing wellhead pressure itself is the result of allocated gas volumes and the production rates, so the solution would be iterative. Gas lift response curves generated for a range of wellhead flowing pressures are required for true optimization.

Fig. 4-12 shows a family of gas lift response curves for a well, generated for flowing wellhead pressures of 100–400 psig in 100 psi steps. The shaded area enclosed between the curves of 100 psig and 400 psig and the minimum and maximum gas lift rate lines (PQ and RS, respectively) is the three-dimensional space used by the gas lift optimization algorithm to find a valid operating point.

Line PQ refers to the minimum amount of lift gas that is required for a stable gas lift operation, considering the operating gas lift injection depth of the well. It is assumed that the well is being gas lifted through a single-point lift at the operating mandrel station, and all unloading valves above are closed. Line RS represents the maximum production rates for various flowing wellhead pressures corresponding to the zero incremental slope; i.e., at that wellhead pressure, increasing the lift gas rate will not lead to any further increase in oil production from the well.

Assuming a gas allocation of 2.5 MMscf/Day for a flowing wellhead pressure of 200 psig, the well would produce 1230 BOPD under these conditions (solution point **S1**).

The optimization program, making use of the gas lift availability in the network, makes several attempts to increase or decrease lift gas to the wells in the network, driven by the economics (i.e., volume of oil and water produced vs. quantity of lift gas utilized). To achieve a solution between the gas lift response curves, it is able to interpolate between the gas lift response curves already generated for specific FWHP values. Considering the combined performance of the wells connected to the entire network, after several iterations, it allocates 2.0 MMscf/Day to the well, for a resultant FWHP of 180 psig (point **S2**, lying on the extrapolated gas lift response curve for this well). In the final solution point, the FWHP and Q_{gi} have decreased, but not the oil production rate.

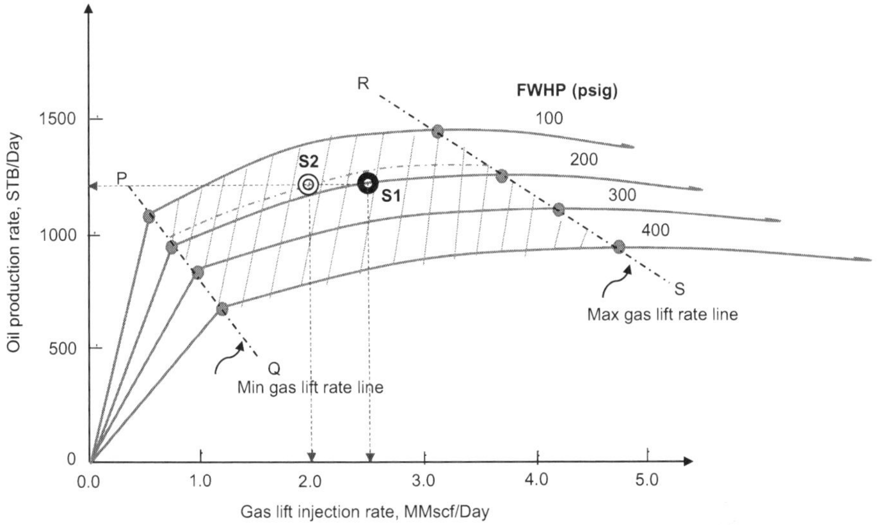

Fig. 4–12. Gas lift optimization by system optimization program

4.4.3 Pipeline De-bottlenecking

Often in existing fields, the pipelines carrying produced fluids to the separation facilities become a bottleneck; this increases the flowing wellhead pressures of the wells. Excessive pressure drop in lift gas distribution pipelines reduces the gas lift supply pressure available at the well location.

If the well production flowlines and pipelines were originally sized for naturally flowing conditions, they may not be adequate when the wells are placed on gas lift. This will create significant back pressures on production platforms that are far away from separation facilities. This problem gets aggravated with an increase of lift gas to the wells with increasing water cuts. Another common reason for this problem is addition of more wells to one trunk line due to development drilling. But gas lift wells perform efficiently if the back pressure is reduced. For this reason, there is always a search to find ways to de-bottleneck the system, especially the production pipelines.

The use of the field network model for pipeline de-bottlenecking will be illustrated with an offshore platform example. **Fig. 4-13** shows an offshore field with six satellite platforms, all feeding to one Facilities Platform. It has been decided to add one more platform, G, as there are no more slots available for drilling in-fill wells in platforms A or D. Already, platform D has the highest production manifold pressure, and the wells at platform D are producing at sub-optimal conditions due to high back pressure.

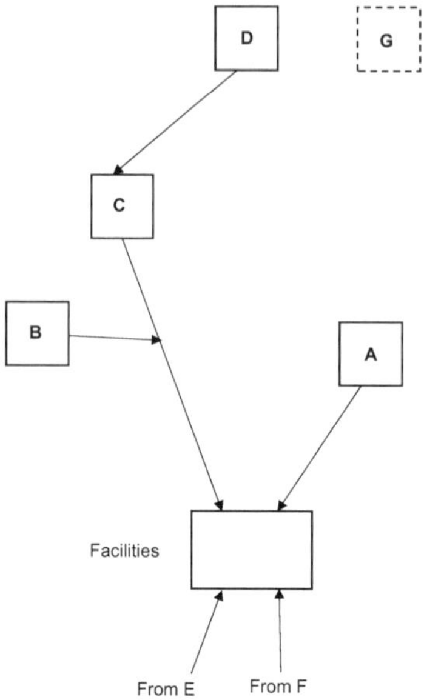

Fig. 4–13. Offshore field with satellite platforms

A long subsea line will require a big investment. A careful investigation of the project economics and viability is required before deciding to install an additional subsea line. The original proposal is to install platform G closer to D and connect the pipelines through a bridge. The production pipeline from G can be brought along the bridge, avoiding a subsea line, but would be connected to the departing pipeline at platform D. This will further increase the production manifold pressure at platform D and affect the gas lifted wells in both D and G.

Using a system network model, several pipeline arrangements with different sizes can be evaluated. The calculation is not only about the pressure drop in a single departing pipeline; it is about the impact it will create on all the wells in the field. Only a system model can answer this question in a technically sound manner. Models will be created for all the wells to be drilled at platform G and will be included in the existing network model, and several options evaluated. A few proposed arrangements for new pipeline(s) are shown in **Fig. 4-14**.

Four options have been mentioned for pipeline configuration for this case, and many more are possible.

1) Run a surface line from platform G to D (original plan)
2) Run a subsea line from platform G to A
3) Run a surface line between platform G and D and a subsea line from platform G to A
4) Run a surface line between platform G and D and a subsea line from platform G to Facilities

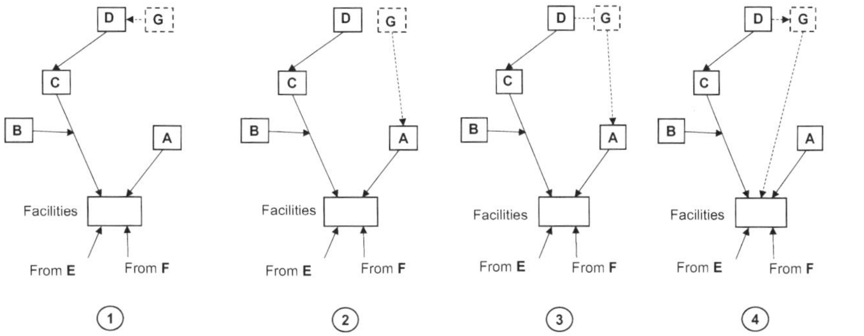

Fig. 4–14. New platform hook-up options

The final flow configuration will also depend upon how loaded the existing pipelines are. The best combination that gives maximum net revenue can be used to justify the pipeline project. The opportunity should be used to ease out the existing bottlenecks in the field. This will improve oil production from the field and manage the lift gas demand efficiently.

The well models can be updated for the expected conditions of years 1, 3, and 5, and the runs can be repeated, if needed, before running economics. Usually, the payout should be within one year, so the current year model should be sufficient to provide the justification.

4.4.4 Field Development Forecast

The typical oil production profile from a field is shown in **Fig. 4-15**. It has three stages: Build-up, Plateau, and Decline. The network model is extensively used in managing the second and third stages of field development.

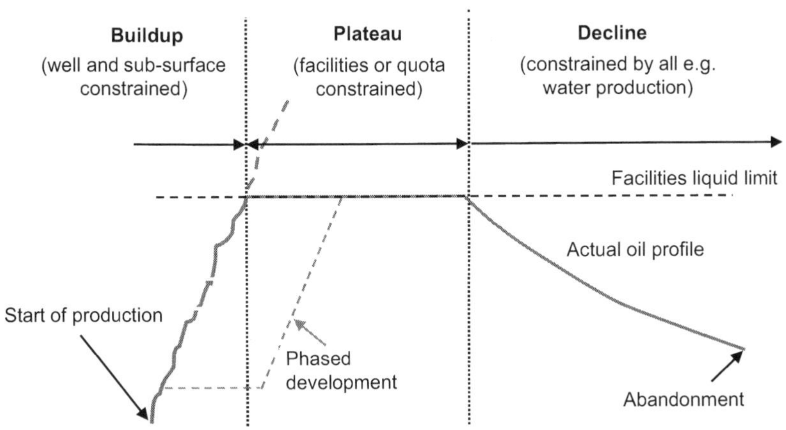

Fig. 4–15. Field oil production profile

A network model can be connected to a reservoir model [either a material balance model or the specific output of a reservoir simulation model, obtained as External profiles (EPF files)] so that it incorporates the constraints imposed by the network while determining the production rates. The reservoir output can be on annual, semi-annual, or quarterly time steps.

This can be more efficiently done with the help of a forecast tool that will manage the integration between the reservoir model, surface network model, and the activities timeline (drilling schedule, equipment commissioning schedule, etc.), to produce a field forecast that can be achieved with these inputs. The reservoir component will predict the change in reservoir conditions over time and will feed to the network model for each step of calculation. The rig schedule and equipment planning will also get factored into the asset forecast.

If, for example, the combined model is run annually, it will incorporate the new wells as per the drilling and workover program and will bring online the additional equipment (pump, compressor, separator, etc.) at the time when it will be ready or needed. The model will run cases for future years, one year at a time, and will predict production rates that are achievable, considering the constraints from wells and the surface network. If needed, it will shut off the high water cut wells. The results are far more reliable than running a reservoir simulation alone. The schematic of such a production forecast is shown in **Fig. 4-16.** [20]

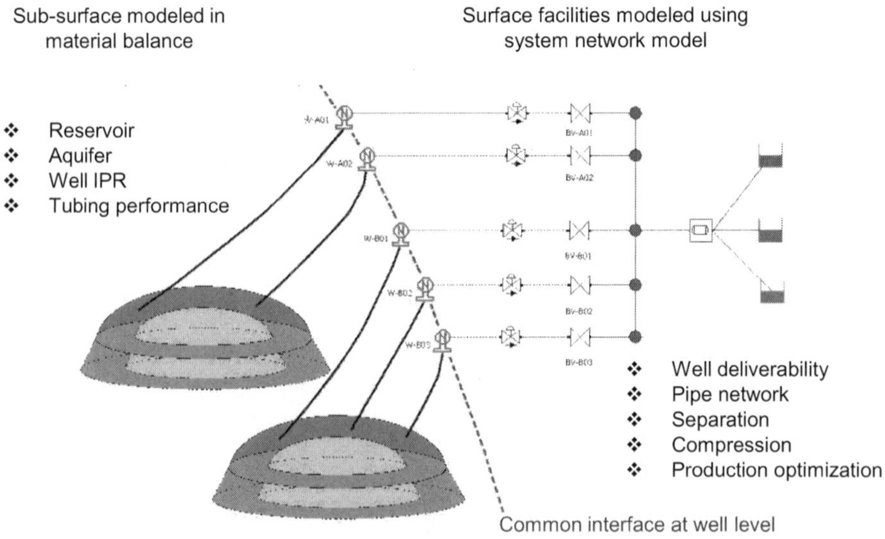

Fig. 4–16. Production forecast schematic

In this process, the network model calculates what the system can deliver at a point in time and also calculates the well rates. Considering the production and injection volumes, the reservoir model calculates how the reservoirs decline with time. The reservoir model also calculates the well performance in terms of production rate, produced GOR, and water cut vs. THP at a given time. It can optimize each time step and generate an optimized profile.

A drilling schedule can be pre-set or wells activated automatically by the optimizer to meet demand. In the forecast sequence, the drilling schedule can respond to a specific, identified network constraint. Downtime can be specified explicitly (e.g., for wells, equipment) or as a period average. In other words, if the target production is set, the optimizer can control the well drilling schedule and field activities and show how the target can be achieved over time. Using the equipment commissioning timeline, design and procurement of several items can be planned in advance.

4.4.5 System Optimization Workflow

The well modeling and calibration becomes an essential part of the network optimization process. This is shown in the form of a flow chart in **Fig. 4-17**.

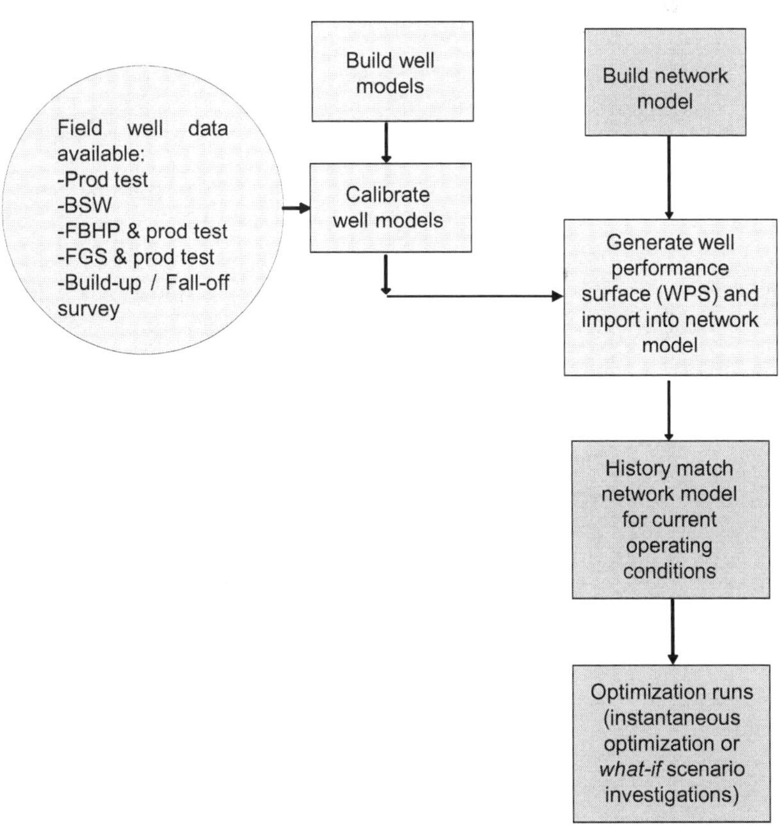

Fig. 4–17. Well model calibration as a part of network optimization

The system optimization cycle of an asset requires performing six distinct activities, as summarized below:

1) The well models are recalibrated using the most recent production test data. The equipment models are re-tuned using actual data from the field. After re-calibration, new performance curves are generated for all the wells and equipment to reflect their current performance capabilities, and these are embedded in the network model.
2) The network model is configured to reflect the status of the field under current conditions:
    a. Status of wells (open / shut-in), process equipment (on / standby), and pipeline flow configuration
    b. Equipment availability or constraints (e.g., not available due to planned maintenance, speed limitations to be adhered to due to current equipment conditions, etc.)
    c. Economic parameters such as concession oil price, fuel gas or power costs, and water disposal costs
3) The model is run in optimization mode. The results are reviewed (many re-runs may be needed), and the recommendations of the optimization model finalized and sent to the field.
4) Operations implement the new set points and monitor the results of implementation.
5) The recommended data is entered into the corporate information systems; the well and network models used for optimization are archived for future reference and use.
6) The field adjusts and stabilizes with the new set points. New field data is collected, and steps 1–5 are repeated again for the next cycle.

This is captured in the workflow schematic shown in **Fig. 4-18**. It is clear that the process requires good coordination between field operations, maintenance, and engineering teams.

As a part of the online optimization, the operator routinely collects field data, updates and calibrates the network model, performs optimization runs, and implements set points in the field results at regular intervals. This becomes a routine in ongoing operations. The operator has a process in place to update the model structure to reflect changes effected in the field. Having established a routine production implementation process, the operator will become fully versatile in the use of the network optimization model. Due to the offline work process, the model lags behind the field by a few days or weeks.

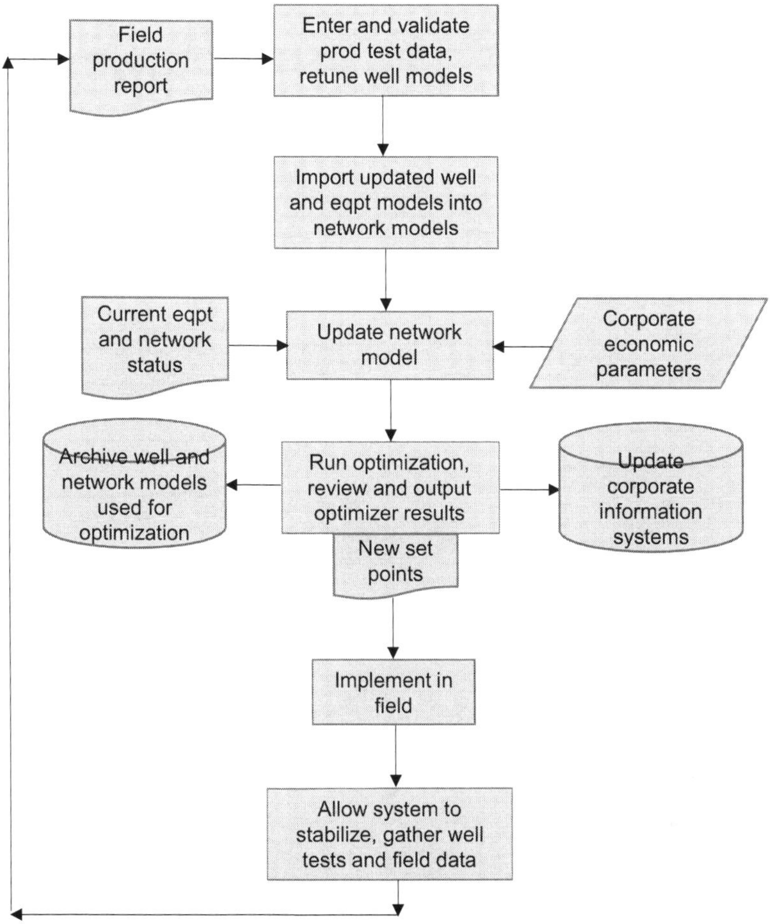

Fig. 4–18. System optimization workflow with data backup

5

System Modeling and Optimization Case Histories

It is worthwhile for operating companies to observe the trends in the industry. This helps the operator know which new technology works well and the level of effort that is needed to achieve success. Having become aware of the potential pitfalls, the operator can take suitable precautions when beginning new initiatives. There is so much sharing of information happening these days through publication of technical papers and conferences, but what is shared in public is usually a highlight of results. The hard work that went into it should be learned from individuals who worked on the project. The lessons learned from any project are equally as important as the results achieved.

In this chapter, a first-hand account of a few case histories of system optimization will be presented.

- Integrated production optimization of an offshore field in Asia
- Production optimization of giant offshore fields, including gas lift compression in the Middle East
- Use of system optimization for field redevelopment

As required by protocols, it is not possible to publish the confidential information of clients. However, the technical aspects covering the challenges encountered, the solutions offered, and the benefits realized from a few total system optimization projects will be presented.

5.1 Production Optimization of Offshore Field in Asia

Project 1 is the development and implementation of an integrated production optimization model for a client in Asia for their offshore operations. The field had 119 production strings at that time, with most of the wells on gas lift. A dual completion well is assumed to have two production strings. Continuous efforts had been made in-house to maximize production by optimizing gas lift on a well-by-well basis. The optimization of the total system was a challenge, in the absence of an integrated production network model. This was a closed loop rotary gas lift system, returning gas lift gas along with produced fluids to the central production facilities. The back pressure on the wells was always a concern. Hence, the operator decided to develop a total system model that would help optimize the wells and network at a system level.

5.1.1 Project Summary

Models were built and calibrated for all the wells in the field. Individual well gas lift diagnostic work was performed during the well modeling phase. Remedial work was implemented in the field simultaneously to improve gas lift performance of wells that were performing sub-optimally. Considerable production uplift was obtained from this work.

The network model was built, history matched, and calibrated to actual field data, within acceptable tolerance limits (<2% for the measured liquids production). Optimization investigations conducted using the total system model resulted in additional oil production gains along with marginal reduction in total lift gas requirement and compressor horsepower.

Successful implementation of the network model required a significant amount of commitment and involvement from the Operations Division. A multi-disciplined team, including a well production engineer, facilities engineer, and well operations personnel, was formed in order to ensure that the changes initiated in the field were well coordinated. The results were carefully monitored, documented, and analyzed at each change.

The project execution involved the following activities:

- The well and network models were built using the data provided by the production operations and engineering teams.
- Wireline interventions were carried out in some wells to correct multi-point gas lift injection and increase gas lift operating depths. This improved the gas lift efficiency of some of the underperforming wells.
- The well models were updated with the new gas lift and production test data, and well performance curves were created and imported into the network model.
- The network model was history matched, and wells, pipelines, and pieces of equipment with sub-optimal performance were identified.
- Optimization runs were performed using the history matched model. Reallocation of lift gas (increase or decrease) was done in step changes, closely monitoring compressor performance and fluid production rates.
- The net gain in oil production achieved from the project was 3.5%, with a simultaneous reduction of the lift gas requirement by 6%.

It was proved that the technology utilized in this project was capable of modeling large-scale, complex production networks in a robust and reliable manner. It was recognized that maintaining the integrated production network model would help the engineers analyze and diagnose a wide variety of production- and equipment-related issues in the field, apart from gas lift system optimization.

5.1.2 Project Steps

The development of an integrated production model comprises several distinct phases, all of which require appropriate software applications, good quality data, and engineering expertise to provide a meaningful solution. These phases are summarized below:

1. Fluid PVT modeling
2. Well model construction and calibration
3. Pipe and flowline modeling
4. Process modeling
5. Model integration
6. History matching
7. Running optimization scenarios
8. Field implementation and follow-up

5.1.3 PVT Modeling

The PVT data available in the asset was reviewed. It was necessary to calibrate the PVT data using appropriate correlations to use at the well and network levels.

Black oil modeling was used as the basis for fluid modeling, using the black oil PVT analysis data that was available with the operator. PVT analysis data of four major fluids was used in the model. A compositional fluid model would not add any particular value to the gas lift optimization of the field.

Since a black oil fluid source group was used throughout the network, it was necessary to describe the PVT behavior of the four major fluid types in terms of a single PVT correlation, so as to allow correct mixing of fluid streams in the network. To address this challenge, a novel tuning method was implemented, wherein two tuning parameters were introduced into each of the available PVT correlation equations (e.g., Standing, Vasquez-Beggs, Glaso, etc.). Each correlation was then "tuned" over the entire four PVT lab datasets, adjusting the two tuning parameters to get the best agreement. The correlation with the best overall "fit" to the PVT lab data, i.e., having the least RMS (root mean square) error in the fit, was selected for the network model.

5.1.4 Well Modeling and Review of Gas Lift Practices

The initial phase consisted of constructing a base well model for each producing string, using the most recent well and reservoir data provided by the operator. Each model was then calibrated to the latest available flowing gradient survey (FGS) data. The calibration was used to identify the most appropriate multiphase flow correlation and tuning factors. If the available FGS data was old (several months to years) and the production conditions had since then changed considerably, the models were updated with the latest production well test data, using the multiphase flow correlation identified earlier and honoring the lift depth (i.e., gas lift mandrel position) identified using the FGS survey.

The well modeling also involved the modeling of the gas lift side of the well. The lift gas flow rate passing through the operating gas lift valve (normally an orifice) is a function of the orifice/port size and the upstream (casing) and downstream (tubing) pressures. Using the lift gas injection pressure and rates measured in the production well test, it is possible to "calibrate" the individual well model's gas injection by adjusting an orifice parameter until the gas injection rate predicted by the model for the measured casing head pressure agrees with the actual measured gas injection rate.

Finally, well performance surfaces (WPS) were generated for each well. These were assigned to the respective wellhead object in the network model.

The gas lift practices being followed in the asset were reviewed, and specific recommendations were made in the following areas for long-term implementation (**Table 5-1**):

Table 5–1. Review of gas lift field practices

S No	Category	Recommendations
1	Well gas lift design changes	• Ensure single-point lift • Improve gas lift depth using the available injection pressure • Update gas lift design for changed inflow conditions • Use orifice valves at operating mandrel station
2	Flowing gradient survey procedures review	• Test the well during FGS • Use electronic memory gauges in place of mechanical gauges • Use better planning for gradient stop depths and timing • Monitor the well closely during FGS
3	Gas lifted well surveillance improvements	• Improve well test frequency • Address gas measurement issues at test separator • Improve gas lift valve reliability by tracking gas lift valve performance and identifying modes of failure (work with gas lift valve supplier).

5.1.5 Network Model Preparation

The field had a process complex consisting of five bridge-connected platforms. The process complex had three-stage separation, pumping, and gas compression facilities. There were thirteen remotely located wellhead platforms at that time.

5.1.5.1 Pipe and Flowline Modeling

A total of 452 pipe elements were eventually included in the model with the following breakdown:

Multiphase flowlines	227
Gas flowlines	165
Risers and downcomers	60

In both the well modeling as well as the riser-downcomer modeling, care was taken to ensure that riser components were configured such that thermal conductivity of the relevant surrounding medium (i.e., air, water) was taken into account. All subsea pipelines were modeled with a 1″ concrete insulation coating.

'Panhandle A' flow correlation was used for gas pipelines and gas injection lines in the model. A modified Beggs & Brill correlation was used for the multiphase / liquid lines.

5.1.5.2 Process Modeling

The role of the process side in the integrated model was to accurately represent the flow paths, fluid movement, and mixing behavior of the production fluids. The main components of the process—the chokes, pressure control / flow control valves, compressors, heat exchangers, separators, flares, and pumps that would directly affect the pressure, temperature, and flow distribution of the production fluids—were included in the model.

Besides the production and injection manifolds on the wellhead platforms, the network model included the following major process elements:

- Four high pressure (HP) production separators
- Vessels for medium, low, and very low pressure (MP, LP and LLP) separation
- Five gas lift compressors
- One recycle loop for LP gas recovery

The production fluid network and lift gas network were part of one model (they were not modeled separately and integrated).

The separator models employ rigorous flash calculations to determine the phase split at the operational temperature and pressure, and separated fluids leave the separator object through different exits. Entrainment of gas in oil or water, oil in gas or water, and water in gas or oil could be modeled. The operational pressure could be fixed (as constraints) in the optimization runs or placed under optimizer control.

The gas compressor was modeled using a thermodynamic relationship between throughput, compression ratio, horsepower, and efficiency. Heat exchangers in the process network were modeled as simple heater/cooler type with specified target temperature.

Finally the model was connected to external source and sink objects to connect the produced fluids coming from adjoining fields to the production from the subject field, with which they were commingled, before being exported to the terminal.

5.1.6 History Matching

The primary purpose of the history match is to establish confidence that the model is representative of the actual producing network, by mapping and comparing the predicted pressure / rate / temperature profile across various points in the network against measured field data.

For a gas lifted system, the history match requires running the model after fixing the production separator pressures and lift gas rates to the wells.

In order to match the measurements provided, the field data from one day's production, on a day with stable operations (with minimal disturbance or downtime), was used as a baseline. Constraints and objectives were defined in the model as follows:

- For each well, the lift gas injection rate was constrained to the measured value, fixing the lift gas injection rate into the well.
- A gas injection header pressure constraint was applied.
- Pressures at each separator were constrained to the measured values.
- Delivery pressure constraints were set at the field battery limit for the oil and HP gas lines.
- Source constraints were set for the other source objects feeding the field.

The network model was run with all of the above constraints in place. Flow rates and pressures at various nodes of the model were compared with the measured data. The L (length tuning) and D (diameter tuning) factors for various pipeline segments were adjusted so that the model was able to reproduce the pressures and flow rates measured in the field. This process required a significant number of model runs.

The history match provided a reasonable agreement between the actual data and the history match model results. It also showed the conditions of various pipelines based on the length tuning and diameter tuning factors. However, there were two issues: first, the total liquid rate matched very well, but the actual oil production rate from the field was slightly lower than the model result due to an increase of field average water cut by about 2% during the period the model was being constructed. Second, there was also a discrepancy between the total gas produced as a sum of the well tests and the calculated total from field mass balance calculations. This was a direct result of some excessively high GORs reported in many production tests and used in the well modeling.

5.1.7 Optimization Scenarios

Once the model had been successfully history matched, it was then possible to use it to run optimization scenarios. A number of these were run. The outcomes are summarized below.

5.1.7.1 Optimization 1: Base Case—Lift Gas Reallocation

The constraints imposed at the wells in the history match phase (the lift gas rate to the wells and the gas lift header pressure) were removed, allowing the optimizer the freedom to determine the optimal solution, i.e., in this case, maximizing oil production at the delivery point. The only constraints retained in the model were the HP production separator pressures. The compressor had enough extra horsepower at its disposal for gas compression as required.

The gas reallocation optimization suggested increase of production from the field by reallocating gas to about half of the platforms. **Fig. 5-1** shows the expected production gains from gas lift reallocation on a platform-by-platform basis.

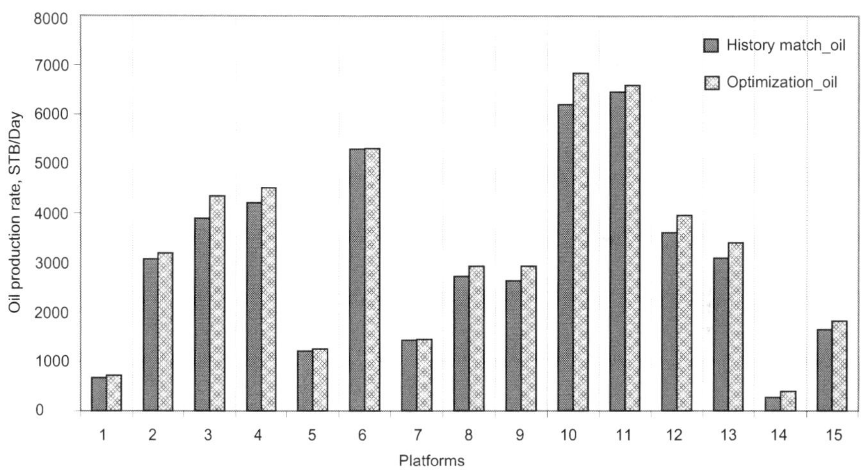

Fig. 5–1. Oil production uplift by gas lift reallocation

Even though the results indicated a possible reduction of about 10% in the lift gas requirement for the gas lifted well network (leading to a horsepower saving of about 14%), it was expected that, in reality, the gas savings might not be as high as predicted. This is because the formation gas-oil ratio (FGOR) value provided for some of the field wells from the well test data was so high that well modeling analysis indicated that those wells may not require any gas lift for production in the first place. Consequently, the optimization algorithm either allocated no gas or made significant reduction in the lift gas rate to those wells. But in the field, these wells needed gas lift to produce.

The model had indicated that there is a potential to reduce lift gas to some of these wells based on network optimization. This, combined with the proper allocation of optimal gas to the wells would yield the increase in production rate.

5.1.7.2 Optimization 2: Separator Pressure

This optimization scenario was investigated by running the model with only the separator pressure being fixed with a bound constraint. The model was left free to decide the optimal lift gas rate for the wells and the gas lift header pressure required for the field.

The model was run for five different separator pressures over the range 6–8 Kg/cm^{2}g, with increments of 0.5 Kg/cm^2. Based on model predictions (**Fig. 5-2**), the increase in oil production rate for every Kg/cm^2 pressure reduction of separator pressure could be estimated. The model also showed the lift gas saving associated with this separator pressure reduction. Considering the design limitations associated with the gas lift compressor, the separator pressure achievable in the field was at 7.0 Kg/cm^{2}g, which was very close to the history match value. Thus, the separator pressure was already at optimum.

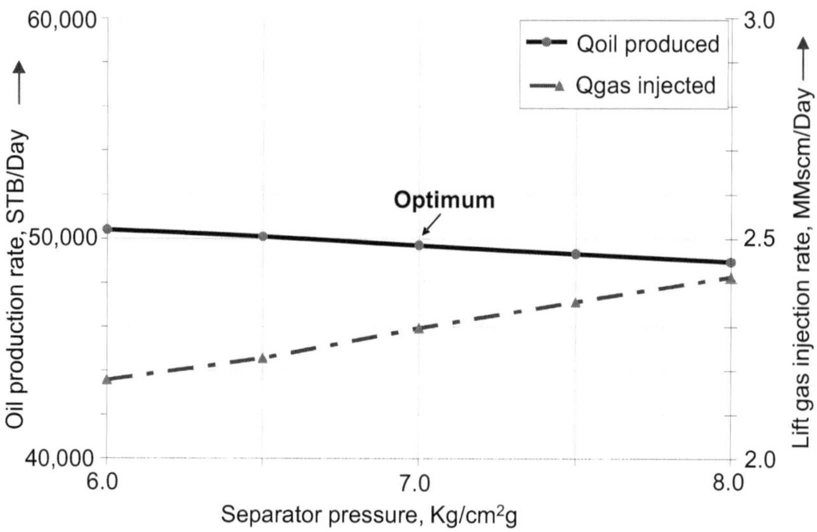

Fig. 5–2. Effect of separator pressure on oil production and lift gas
Note: 1 Kg/cm^2 = 14.22 psi or 98.067 kPa.

5.1.7.3 Optimization 3: Gas Lift Header Pressure

Another scenario investigated was to run the model for various gas lift header pressure settings, keeping the separator pressures unchanged at their history match values. Results of this confirmed that the current gas lift header pressure was at optimum for the field gas lift network, and increase in gas lift header pressure would not result in any additional production from the wells. The extra gas pressure would be only absorbed as pressure drop at the well gas injection chokes, hence not value-adding.

5.1.7.4 Optimization 4: Pipeline De-bottlenecking

A 10-km-long departing pipeline from one of the satellite platforms carrying produced fluids to the facilities platform was selected as a potential candidate for a pipeline de-bottlenecking scenario. This was one of the pipelines that had the maximum gradient of pressure loss. The nominal size of the pipeline carrying well fluids was increased from 10″ to 14″ and the model was run at the same separator pressure used for history match.

The results indicated that this increased pipeline size would decrease the back pressure on the production manifold by about 4 Kg/cm^2 and would increase the production from platform wells by about 450 BOPD (**Table 5-2**). However, the decision to upsize this pipeline can be made only after studying the economic recovery of this investment—and after ascertaining the improvement in the recovery of reserves and the magnitude of the acceleration component of production.

Table 5–2. Pipeline de-bottlenecking study

	Platform oil production rate, STB/Day	Platform water production rate, STB/Day	Platform production manifold pressure, Kg/cm^2g
Existing pipe size 10″ Nominal	4347	2261	13.87
Proposed pipe size 14″ Nominal	4792	2512	9.08

It is worthwhile to consider other options such as using redundant lines available in the field to reduce the back pressure on some of these production manifolds. Where possible, production from a few wells could be diverted through an alternate pipeline, and this would reduce the production manifold pressure on the platform. The current optimization model could be used to evaluate the benefits (possible oil production uplift and/or lift gas reduction) of such alternative modes of operation.

5.1.8 Field Implementation

After completing the optimization scenarios and a detailed review of the results, the operator commenced a program to implement the recommendations.

The agreed plan was to implement all the field-wide changes together, but in small steps, if required. If the gas lift rate has to be reduced by 10%, the change may be done in a single step. But if the decrease is 25%, it should be done in two steps: change the parameter, allow it to stabilize, monitor, and review the results. If the results are in the right direction, continue to the next step. If the result is not good, review and update the model.

It was also clear that measurable changes in system production are better monitored at the facilities level. The individual production well tests are more useful for well performance review and well model calibration.

The whole implementation process required all personnel involved to work as one team. The implementation in this case was done at two levels:

- Well interventions on the most promising candidates identified during the well modeling segment of the project. These consisted primarily of gas lift valve change-outs and changing point of injection. Unfortunately, due to mechanical reasons, not all the candidate wells could be re-entered. Some wells required additional workover.
- Adjustment of lift gas injection rates in line with the base case gas lift optimization results.

The benefits obtained from some of the individual well gas lift optimization are shown in **Fig. 5-3**.

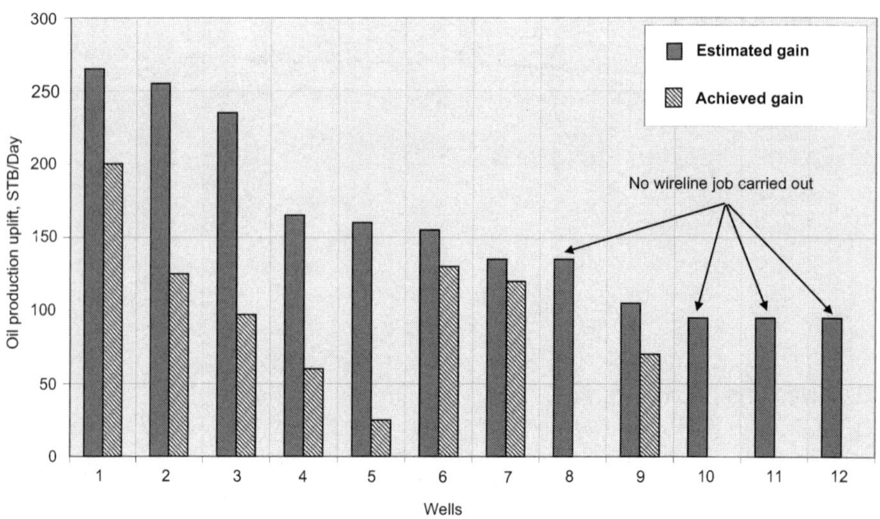

Fig. 5–3. Well gas lift optimization results

The production gains realized from the field-wide gas lift reallocation are shown in **Fig. 5-4**.

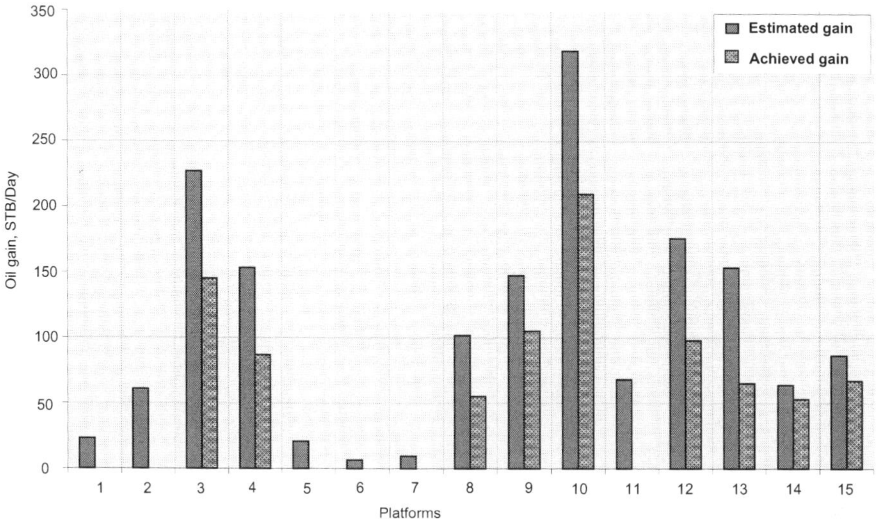

Fig. 5–4. Network gas lift optimization results

The combined benefits of these two steps yielded an increase of about 3.5% in the oil production rate from the field, along with a 6% reduction in the lift gas requirement.

The gains obtained by any system optimization will taper off over a period of time, as production conditions and well performance change. In order to obtain continuous benefit from the integrated production network model, it will be necessary to keep the model up to date and to re-optimize the network at regular intervals. This will ensure the system is performing at optimum levels at all times. This can be done by re-tuning the well models with the latest production test data and replacing the existing WPS with new ones. Any change of flow lineup or well or equipment status should also be captured in the network model.

5.1.9 Project Conclusions

1) An integrated production model of the field was successfully constructed, history matched, and used to optimize lift gas allocation for increasing oil production. The technology provided by the service company was robust and reliable.
2) Optimization results predicted a potential uplift of 6% in oil production. However, actual oil production uplift achieved was about 3.5% of total oil volume, along with a saving of 6% of lift gas volumes.
3) The study confirmed that the current gas lift header pressure and production separator pressure settings were close to optimum for the field. The model indicated possible production loss if the separator pressure was increased.

4) No significant change in oil production was predicted as a result of changing the lift gas header pressures.
5) Potential gains may be realized by increasing the diameter of some of the trunk fluid pipelines; however, implementation of such recommendations requires economic analysis of the investment required.
6) Some of the wells require additional flowing gradient surveys to be run in order to recalibrate the well models, as some of the current well models are based on old and outdated data. Updated modeling of these wells will improve the reliability of the network model results.
7) The gas measurement from the test separators should be improved to get actual produced GOR from the production well tests. This would improve the well model calibration.

5.2 Production Optimization of Major GL Fields in the Middle East

5.2.1 Project Background

Project 2 is the implementation of a total system production optimization model in a complex offshore operation in the Middle East.

The Middle East operator operates four fields offshore, with gas lift as the major form of artificial lift. The primary reservoirs are mature, water-flooded carbonates. Two main fields house the central separation and compression facilities. Lift gas is distributed to the satellite platforms and the smaller fields via independent pipelines. A common lift gas pipeline allows gas transfer between the two main fields. A total of fourteen major gas lift compressor trains provide the required gas lift compression. These are large-capacity multi-stage centrifugal compressors driven by gas turbines similar to the engines used in the aviation industry. Oil, gas, and water produced from the platforms is returned via pipelines and processed at the central facilities. **Fig. 5-5** presents a simplified schematic highlighting the oil and gas interaction between the fields.

The project involved optimization of the asset through integrated asset models containing:

- More than 200 gas lifted wells located on 40 satellite production platforms, spread over four offshore fields
- Associated flowline and pipeline networks with valves, heat exchangers, pumps, and other process equipment for gathering production and distributing lift gas
- Central gas compression and fluid separation facilities

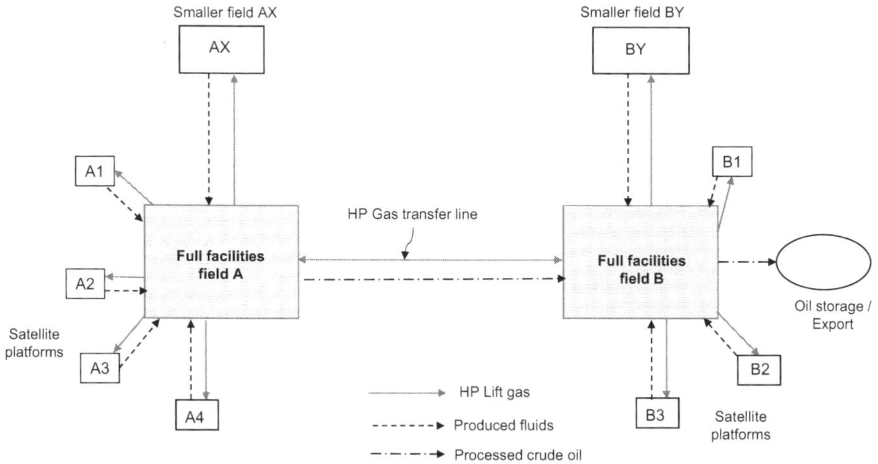

Fig. 5–5. Interaction between offshore fields

The gas lift operation in the asset was horse-power constrained. One of the major expenditures for the company was for maintenance of the compressor trains to operate these units at peak efficiency. Engine overhauls were a routine operation to regain horsepower of the gas turbine engines. From the time of overhaul, the GT performance would deteriorate as the machines operate and would reach a point when it would require overhaul again. Moreover, the available horsepower would be affected by the ambient temperature and the wind direction. In the Middle East and Asian regions, as the ambient temperatures increase from 75 to 95°F from winter to summer, the horsepower output from gas turbines decreases by about 10–12%. This seriously affects the horsepower available.

On extremely hot days when the available horsepower was much less, the company used to run a few of the gas turbines above based load (ABL), i.e., at speeds slightly higher than the designed speed of the machine. The ABL operation was carried out with additional surveillance in place. The disadvantage of this operation is that, for those engines, even if the engine is run at 4% ABL for a few days, the time between overhauls (TBO) gets reduced, and the engine needs to be sent for overhaul sooner. This, in turn, drives the maintenance cost up and also causes more production downtime.

The objective of the implementation of the total system production optimization model in this case was to *increase production* and *reduce operating cost.* The asset already had optimization models that considered the wells and production gathering network, but not the gas lift side of the operation. The new model, therefore, needed to be able to consider the combined performance of the total system, including downhole well configurations, the complex production gathering and lift-gas distribution pipeline networks, separators, compressors, and pumps. This was particularly important as the compressors were a constraint on total system performance in the fields.

In this case, the model needed to be able to:

- Determine suction and discharge pressures of the gas compressor and the gas lift rates optimally needed for the entire production system, and determine the *horsepower requirement* for the full network.
- Determine the *horsepower available* from the detailed modeling of the compressor drives (i.e., gas turbines).
- Provide an optimal solution based on the above two considerations.

When generating the solution, if the horsepower available was limited, the model could run all units at maximum horsepower; but if there was excess horsepower, it could slow down the machines or keep one of the units on standby. Being able to keep a train standby might seem to be a luxury, but if it is possible to do it, it is great because the standby machine can be brought online during the planned or unplanned maintenance of a compressor. This can save lost production. In the industry, machines running at speeds close to 10,000 rpm will always have some minor issues (vibration problems, seal oil leak, etc.) so shutting down these units to rectify these problems will be required every now and then.

Apart from set points on gas lift supply pressures, well gas lift rates, and production separator pressures, the model needed to also provide guidelines on the gas sharing between the two fields in a productive manner that was advantageous to the corporation. Due to the difference in field average water cut and FGOR of the wells, the amount of associated gas produced in each field was different. The installed compression capacities were also different. There was only one gas transfer line so it was used for sending gas in only one direction at any given point in time. As the low pressure gas was not returned back to the circulation loop, the amount of gas taken away from a field was limited by the associated gas produced in the field.

5.2.2 Model Preparation

The system optimization solution that was provided met the required needs, as mentioned below:

- Using the advanced gas lift valve modeling feature of the well modeling software, it was possible to model each well's gas lift parameters and study the effect of varying injection pressure, gas lift flow rate, and the operating gas lift valve port-size on well performance. The gas lift injection side was fully included in the well performance analysis when calculating the well performance curves of the gas lifted well.
- The compressors were modeled using a detailed compressor object that would allow the estimation of available compression horsepower under existing ambient and field conditions on any day.

Presented below is a summary of the major activities performed during the development of the production optimization model.

5.2.2.1 QA/QC of Data Received

The data was checked for accuracy and consistency. In general, the data quality was good. Where the data was inconsistent, discussions occurred between the operator and the contractor, and a solution was mutually agreed upon. For example, the gas compressor flow measurements were not in agreement with the gas lift injection rates measured at wells, and there was a difference of about 5-8%. For optimization, the well gas lift rates are used by operations, so it was decided to use the current well gas lift rates as the baseline data, and the compressor discharge flow rates were adjusted accordingly.

5.2.2.2 Construction and Tuning of PVT Models

Black oil fluid correlations were used for the full-field network models, as this is more appropriate for gas lift optimization.

Various fluid models representing fluids flowing from different formations were first tuned with respective experimental data. These PVT models were used in the well models. Each wellhead object was assigned to the fluid model corresponding to the formation the well was producing from.

A single set of correlations suitable for the network was selected and tuned. This was used for the network.

5.2.2.3 Well Model Preparation

The operator used gas lift designs involving an orifice valve for the operating mandrel station and bellows-charged unloading valves in the stations above. Flowing gradient surveys were available for most of the wells, confirming single-point gas lift injection. The existing well models were calibrated for the production and gas lift parameters. For the wells planned to come on production during the execution of the project, well models were constructed. This involved fluid modeling, detailed modeling of inflow and outflow descriptions, including a gas lift design for every new well.

5.2.2.4 Compressor Model Preparation

The detailed compressor models were built using the performance test data of the gas turbines and compressor stages in service. The temperature correction factors for the gas turbines were incorporated. The compressor sections were modeled using the performance curves for each compressor stage provided by the compressor manufacturer, including surge and stonewall limits. The model was completed by incorporating other components such as anti-surge loop, scrubber, and heat exchangers. The compressor models were calibrated with field data.

5.2.2.5 Construction of Network Model

The network model was constructed using the pipeline and equipment data provided by the operator. The wells, gas lift distribution, and production gathering systems were modeled. The satellite production platforms and the central processing platforms (housing the multi-stage fluid separation and gas compression trains) were modeled in detail. All critical components such as heat exchangers, pumps, compressors, and separators were modeled. Block valves and control valves were included in the model. The inter-field pipelines for the flow of gas and produced fluids were incorporated.

The network model was first built with simple compressor models and checked for proper operation. The detailed compressor models were constructed separately and tested. Later, the simple compressors were replaced with detailed compressor models in the network model. This approach allowed thorough investigation of the compressor models before they became a part of the network model, saving modeling time.

5.2.3 Model History Match and Optimization

5.2.3.1 Network Model Calibration and History Matching

In order to calibrate the model to the current field operating conditions, the model was run with constraints set to represent actual production separator pressures and well gas lift injection rates. As the well models had already been tuned, the model production was very close to actual field production. The L-factors of various pipelines were adjusted (effectively adjusting the lengths of the pipelines) in order that the model pressures at various nodes matched reasonably well with the measured pressures from the field. The network model was history matched.

5.2.3.2 Optimization Runs

Except the lower and upper bound pressure constraints on the pressure range of the high pressure production separators, all constraints in the history matched model were removed. When performing an optimization run, the model had the freedom to choose optimum separator pressure, optimum gas lift header pressure, and optimum well gas lift injection rates, in order to maximize oil production within the system of the available field equipment and piping network. The network model ran iterations with the aim of maximizing the net revenue and returned a solution for the total network.

Two large models were created for the four fields (aligned with field operations): one model for fields A and AX and another model for fields B and BY. In both cases, the following optimization scenarios ("snapshots") were investigated:

- Base case optimization
- Determination of the minimum gas lift header pressure required for field network
- One gas lift compressor train shutdown scenario
- Evaluation of optimum gas transfer between fields

A detailed analysis was performed on the results for each snapshot. The results and specific recommendations were presented to the operator, highlighting changes that could usher in significant value.

5.2.4 Results and Implementation

5.2.4.1 Summary of Model Results

1. The model suggested that the system was operating very close to its economic limit with the existing lift gas rate, and there was no significant benefit in adding additional lift gas to the system.
2. The model proposed a reduction of about 5 psi in the HP production separator pressure and a reduction of about 50 psi in the gas lift header pressure for the field. Field implementation showed that it was possible to reduce the pressures in line with the model recommendations.
3. In field B, the model suggested that, during the winter months, one of the gas lift compressors could be shut down and put on standby mode, adding operational flexibility. This was a dream come true.
4. For the one gas lift compressor shut-down scenario, the model distributed the available lift gas to all wells, rather than shutting-in any of the wells. The production loss predicted by the model was much lower than the loss experienced by the field. This required additional investigation.

   Investigation pointed to two main contributing factors for this discrepancy. The first factor was that the results given by the model were for stabilized (steady state) conditions, and understandably, the loss would be more during the transition period. The second factor was the "flat" nature of the well performance curves for some of the wells in the normal operating region. This caused less production loss in the model than actual loss in the field for reduction of lift gas to wells. A detailed review of well models resulted in changes to the outflow correlation for the wells in question. **Fig. 5-6** shows the gas lift performance curves of a well, generated by two different flow correlations.

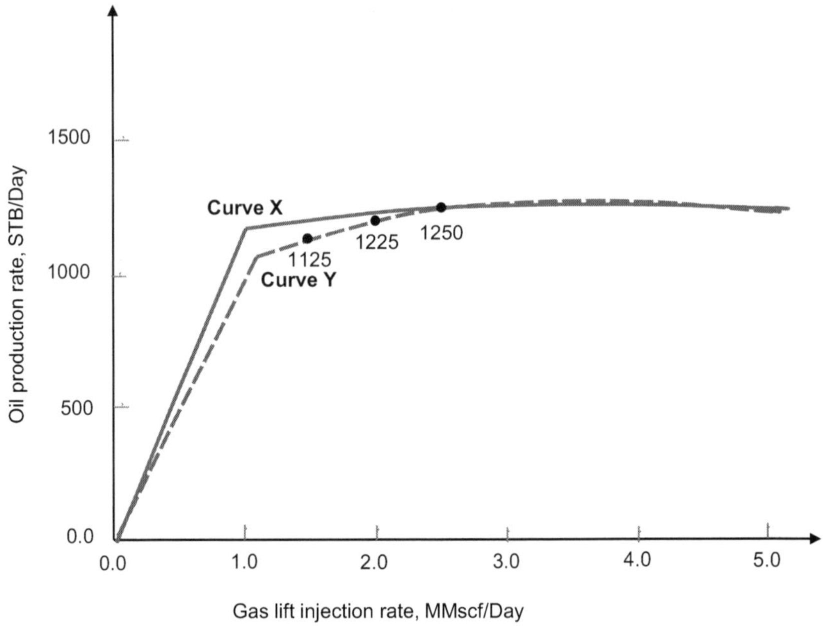

Fig. 5–6. Gas lift allocation with flat performance curve

The well performance curve X is relatively flat above the kick-off point. So when the gas availability is less, the optimizer tends to reduce the gas lift rate, as the loss in production is almost negligible. In reality, if the well performance is as per curve Y, the loss of oil associated with the reduction of lift gas will be much more. With a lift gas reduction from 2.5 to 1.5 MMscf/Day, curve X predicts a loss of about 25 BOPD, whereas curve Y predicts a loss of about 125 BOPD. This reinforces the fact that a well model should be calibrated with a correlation that would better reflect the real gas lift performance of the well under the current lift and water cut conditions. In this case, the optimizer can do a better job of gas lift allocation.

5.2.4.2 Field Implementation

Support from field operations in the implementation of model results was seen as a critical factor for success. Spreadsheet tools were developed to extract additional results from the model. These were useful for tracking the changes made to the wells and actual field results. Key parameters captured during implementation were total fluid production, well test information, separator and gas manifold operating pressures.

Model recommendations called for changes to be made in gas lift header pressure, production separator pressure, and individual well gas lift injection rates. When the pressure changes were effected in a smooth and gradual manner, changes in the gas lift rate set points of the wells (ΔQ_{gi}) were made in steps of

0.5 MMscf/Day at a time. After stabilization, a well test was performed prior to continuing with further change in lift gas rate. If the reported oil loss was more than 50 BOPD for the first change of 0.5 MMscf/Day, the gas lift rate would be adjusted back to the previous rate and the individual well model would be flagged for further investigation.

The model re-tuning is a critical step in the process because it ensures the accuracy of the overall recommendations. This is also to be expected since even slight changes in water cut can affect the model.

5.2.4.3 Frequency of Offline Optimization

Considering the size of the asset and sequence of events, an optimization cycle of three months was used by the asset for offline optimization. While it took about a month for the field to stabilize with the new set points, it took about another month to get at least two stabilized production tests in all the wells. Downhole surveys were performed in some wells, based on an engineering request. With all the data in, it took about a month for the engineers to re-tune all the well models and import new performance curves for all the wells, compressors, and other pieces of equipment and update the settings in the network model, as needed. This allowed a quarterly optimization of the system.

5.2.5 Workflow Improvements

5.2.5.1 Production and Gas Lift Benefits

The project was executed in three stages:

Phase 1—System optimization of Field AX (pilot project)

Phase 2—System optimization of the other three fields

Phase 3—Real-time optimization (implemented later)

The following benefits were realized after implementation of the system optimization model results:

- **Phase 1**

  Field AX: 3% increase in oil production and 14% reduction in lift gas requirements. Reducing gas injection rates to most of the wells helped reduce the separator pressure and resulted in the increase of liquid production.

- **Phase 2**

  Field A and Field B: Increase in oil production achieved was 1% and 0.7%, respectively; the reason for lower increase in production is that these are mature fields with high water cut. The results suggest that the oil production gains could be very high if the water cuts are lower.

Other benefits were the following:

- Achieved 2.7% reduction in overall gas lift requirements. The lift gas that was freed up was reallocated and was used to reactivate two shut-in wells.
- Compressor train operating costs were reduced by 3% on an annual basis. This was achieved via improved compressor management, reduced compressor servicing costs, extended planned preventive maintenance, reduced ABL operations, and reduction in consumption of fuel gas.
- Due to the reduction in compressor suction and discharge pressures, the overall effect was that the compression ratio remained almost unchanged and the gas flow rate was slightly reduced, so the horsepower requirement came down. The reduction in separator pressure reduced the back pressures slightly, contributing to an increase in well production rates.

The benefits achieved due to the system optimization model in this field may seem to be *small in terms of percentages,* but considering the size of the operations, these are very attractive numbers in terms of real values.

5.2.5.2 Gas Lift Reallocation during Compressor Train Down

When compressor trains are shut down for planned maintenance or due to an unscheduled event, there is normally insufficient backup capacity available to keep the system operating at the same parameters. The historical solution has been to shut off gas lift to producing wells in order to maintain a specified gas lift manifold pressure. The wells are prioritized for shutting in, based on the amount of oil production per volume of gas injected. The resulting "hit list" wells are the ones with the lowest gas lift efficiency.

Network optimization runs indicated that reallocating lift gas to all wells in order to maintain the gas manifold pressure resulted in less lost production. This method also considers gas lift efficiency but takes advantage of the fact that the gas lift efficiency gradient is small close to the optimum gas lift rate. Therefore, the overall reduction in oil rate is lower when a small reduction in lift gas is spread over many wells versus shutting in the requisite number of least efficient wells.

5.2.5.3 Compressor Outage

During a compressor outage, operations have the option of redistributing the available lift gas to all the wells instead of shutting in a few wells. This could be accomplished only if the gas redistribution is done immediately, following the shutdown of the train. This could not be achieved in the field, with the current SCADA system and pneumatic-driven automatic gas lift injection chokes. The operator therefore decided to install quick-reacting control valves that would allow the change of injection rate set points to a select group of wells almost instantaneously.

However, the reallocation of gas to the wells is considered to be a better option only if the compressor shutdown lasts for a longer duration. This is because the field as a whole requires some time to reach a pseudo-steady state. If the compressor is down only for a few hours, it would be preferable to shut in the hit-list wells and reopen them once the train is back on line. This would be better than changing gas lift rate set points for many wells in the field. Based on production optimization model runs, look-up tables were created and the reallocation implemented for prolonged shutdowns.

5.2.5.4 Strategy for Planned Compressor Shutdowns

The results of the system optimization model were successfully used to evolve a gas lift strategy for planned compressor shutdowns. The current strategy is to reduce gas lift injection rates of certain wells (as recommended by the network model) in steps in order to minimize the loss of production. This approach is used for planned compressor maintenance events that would last for a few days.

Once the quarterly network model updates are completed, offline network model runs are made in advance for several scenarios of incremental lift gas shortages. The results are documented in spreadsheets and used by the operations personnel to minimize production loss while handling the lift gas shortfall. This approach of using the results of the model for different gas lift shortages has led to a new philosophy where shutting in a well is not the first step for managing lift gas shortage. With the help of the fast-acting gas lift controllers installed in many of the wells selected for gas injection rate reduction, it was possible to achieve the gas lift rate reductions almost instantaneously. This can also be reversed in the same manner. The process works as follows:

- Gas lift rates are reduced by 0.5 MMscf/D per well in a group of 20 wells, and this gives a gas "saving" of 10 MMscf/D. These are called "First Stage" wells.
- If the net lift gas shortage is more than 10 MMscf/D, operations will proceed to reduce gas lift rates in "Second Stage" wells—this will be another round of reduction at 0.5 MMscf/D per well in 20 wells (some of these wells may have been included in the first stage reduction).
- If the lift gas shortage is more than 20 MMscf/D, then the reduction will continue to the "Third Stage" wells and so on.

The production loss due to gas lift rate reduction in first stage wells is very minimal. The loss increases as operations move to the second and third stages, etc. If the lift gas shortage is significant, probably due to more than one compressor shutdown, then it would also become necessary to shut in a few wells (based on gas lift efficiency) from the top of the well "hit list."

The benefit of this approach was that the production loss was kept to the barest minimum at every level, and operations had the flexibility to use the available lift gas volumes under dynamic field conditions. Efforts were taken up to program these groups of wells into SCADA control so that effecting the gas lift reduction could be implemented by simply pressing one button for one stage.

The wells included in each stage might change based on the latest well performance, as captured in the quarterly network model updates.

5.2.5.5 Better Teamwork

The network model implementation and data management process resulted in better communication among those involved in field production optimization. A free flow of communication happened among operations staff, production and process engineers, compressor engineers, and maintenance and reliability specialists, and this enabled them to contribute their best to maintain optimum production under dynamic operating conditions.

5.2.5.6 Creation of Field Optimization Team

A Field Optimization Team was formally created. This consisted of personnel from Operations (Wells and Facilities) and Engineering (Optimization and Production Engineers and Equipment Reliability). They coordinate, monitor, and resolve issues regarding separator and system pressures, compressor operations, gas balances and allocations, and operating scenarios for contingencies. This team worked together to develop strategies for various production upsets. The goal was to sustain a high level of operating efficiency at all times.

A process flow chart was developed to ensure multi-disciplinary input and the support required for managing the network model. This is shown in **Fig. 5-7**. In order to ensure no misunderstanding would occur regarding inputs and data management, a RACI chart (Responsible, Accountable, Consult, and Inform) was created to clarify individual roles and responsibilities.

5.2.5.7 Internal Process Review

The operating company used this opportunity for reviewing its internal work processes. For example, the optimization project highlighted the value of tuning well models and keeping them up to date. This was important for the operator because the water cuts of some of the wells were changing very fast. If the model should match the field production numbers, the well models should be recalibrated at least once a month. The system optimization process gave an opportunity to engineers based in the office to understand operational constraints and bottlenecks that existed at the site but were never highlighted. These issues could be understood and resolved using cost-effective approaches now that they had been raised.

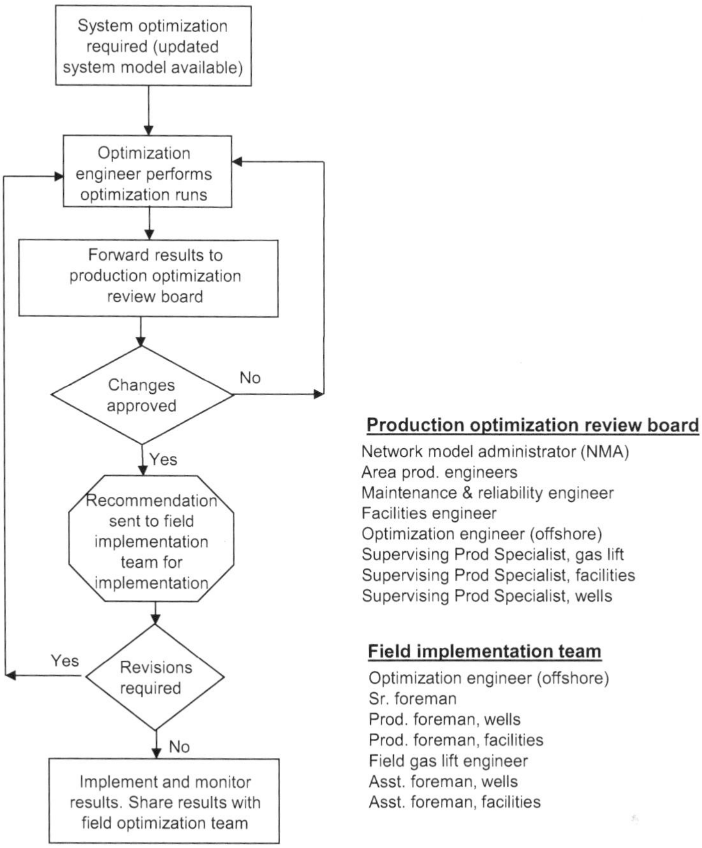

Fig. 5–7. System optimization functional flow chart

The proven benefit of more frequent optimization justified the value of the implementation of full field real-time optimization.

5.2.6 Project Conclusions

1. Implementation of a full field optimization software package resulted in production gains and operating cost reductions. The network model accurately modeled the complex production networks and was history matched across the operating range.
2. Detailed modeling of gas lift compressors provided a definite advantage for achieving total system optimization of gas lifted networks. This was particularly beneficial as the system was gas lift horsepower-constrained.
3. A Field Optimization Team was formed to ensure that all personnel responsible for optimizing the complex and dynamic systems were

proactively engaged. This team would be responsible for maintaining the field optimization strategy, including identification of upcoming system upsets.

4. With the updated network model in place, the frequency of lift gas reallocations using an offline model increased from two to four times a year.
5. During the project execution, it was realized that the engineering manpower required to complete the project had been underestimated. More time and effort were required for calibrating the network model prior to making optimization runs.
6. Production optimization should be seen as a process, not as a one-time project. In order to sustain the optimization benefits, the field network model should be updated on a regular basis and optimization runs performed and implemented in the field. Recognizing the value of this continuous process, the operator reorganized the workflow within their operations. Without this being done, it would not be possible to achieve the benefits of sustained optimization.
7. The outstanding commitment shown by the staff and management of the operating company and the service provider was largely responsible for the level of success achieved in this project. Technical discussions and model review meetings held between the engineers of both companies were mutually beneficial.

5.2.7 Lessons Learned

1. Accurate well model tuning is critical across the full operating range. Well models must be updated frequently to ensure that the WPS represents actual well behavior. This is critical to the accuracy of the model output and program credibility.
2. Detailed modeling of all components may not be necessary to achieve the optimization objectives. This should be reviewed up front before spending time on modeling.
3. The performance curves generated by the well modeling software must be tuned to a wide range of operating conditions in order for the network model to accurately predict operating conditions across the full range of possibilities.
4. It is very important to calibrate the well models frequently and ensure that the "curvature" of the WPS in the model represents the actual well behavior in the field, as software uses these WPS for its calculations.

5.3 System Optimization Model for Field Redevelopment

5.3.1 Background of Redevelopment

Project 3 shows how a network optimization model was used to select the most economic redevelopment option for an existing offshore oilfield in the Middle East region.

The reservoir was a tight and thin carbonate reservoir, developed in the eighties. Even though the field was more suitable for horizontal wells, the wells were completed as vertical producers (horizontal drilling was just getting popular at that time in the region). All the wells were placed on gas lift. The wells would typically come online with a production of about 700–800 BOPD and decline to 300–400 BOPD in a few months. Some wells even declined to lower production rates.

The asset was developed several kilometers away from the central facilities of a major field. There were four platforms in this field, and this platform was at the farthest distance (22 km) from the central facilities. Designed as an unmanned platform, the platform had no living quarters.

The platform was completed with a production separator, but with no main oil pump. Only three pipelines were installed to connect the subject platform to the central facilities of the major field:

- Incoming pipeline for lift gas supply
- Departing pipeline for the separated liquid
- Departing pipeline for the separated gas

A few years later, an incoming water-flood water line was added for water injection wells.

A redevelopment was considered for this reservoir in the nineties, and wells were sidetracked to horizontal holes. This was a major success, and a 3- to 4-fold increase in oil production was observed. Most of the wells were recompleted with horizontal holes. Horizontal wells were added for water injection support as well.

As the water cut increased, lift gas requirement for the wells increased gradually over a period of time. After some years, the problem was very different:

- The available lift gas manifold pressure was about 250 psi lower than the original design
- The production manifold pressure was about 100 psi more than the original design

This was mainly because the volumes of lift gas and the produced fluids handled were more than double what the system was designed for. The gas lift manifold pressure decreased due to excessive gas flow rate in the incoming pipeline. Due to reduced gas lift arrival pressure, the operating gas lift depths had to be shifted to shallow depths in some wells, compromising production. The production separator pressure was the driving force for pushing the liquid and gas into respective departing pipelines from the platform. Due to the increased operating pressure of the production separator, the production manifold pressure increased and the wells were experiencing high back pressure. This affected the gas lift efficiency of the wells on the platform. The problem was so severe that a small increase of lift gas rate to one well would reduce the gas lift manifold pressure seriously, and this would affect production from all other wells on the platform. It was a vicious circle.

5.3.2 Alternatives for Redevelopment

The options explored included traditional and new technology solutions (the author served as the optimization consultant for this evaluation and a subsequent field optimization after installation and commissioning of ESPs in two wells). These solutions are listed below:

1. Run new, large subsea pipelines for the gas lift and LP gas return lines. This would increase the gas lift manifold pressure and reduce the production separator pressure. However, this solution would be the most expensive option. Due to marginal economics and uncertainties, this could not be progressed.
2. Install a main oil pump for the liquid outlet and a pipeline compressor for the gas outlet of the separator. This would allow the production separator pressure to be operated at comfortably lower pressures. This would be more economic than running larger subsea pipelines; however, installing compressor for export was not viable due to deck space limitations on the platform. Another reason was long lead time for a new compressor. The economics were also not very attractive.
3. Install a multi-phase booster pump (MPP) and line up production from some of the wells to the MPP. The discharge of the MPP could be lined up to the separator (or to the liquid departing pipeline). The flowing wellhead pressures of the wells diverted through the multi-phase pump would decrease, and fluid production from these wells should increase. But the overall system bottlenecks would continue to remain almost at the same level.

   Detailed investigation showed that the MPP, if installed, should have the capacity to handle very large gas-void fractions (GVF). With the decrease of wellhead flowing pressure, the GVF of the gas lifted wells—already around 95%—would increase further. This would require a very large pump to achieve a reasonable increase in oil production. Additional deck space would be required. The power generation capability would have to be increased. A special construction barge would be required to install the MPP as the platform crane cannot lift the heavy load of the MPP.

Overall, the economics for the MPP option were positive but not very robust. Hence this option was not pursued.

4. Convert two of the big producers from gas lift to ESP lift. This would reduce the gas lift requirement of the platform, and the gas lift manifold pressure would increase. The reduced gas load would also help reduce the production manifold pressure. Very minimal surface modification would be required. But this option needs high quality power, suitable for ESP.

A few more options considered are listed below. These could not provide a permanent solution to the problem:

- The use of surface ejector pumps with water-flood water as power fluid was considered. This might reduce the FWHP of a few wells, but would overload the separator, as the spent power fluid would also go into the separator along with the produced fluids.
- The use of a pipeline drag reducer agent was tried in the liquid transport line, but the result was not promising.

For any of these options, apart from the benefits of the method selected, the downtime required to execute the work offshore is a significant factor to consider. If the platform produces 20,000 BOPD and if the production has to be shut in for 24 hours to do modification work (especially hot work) or for a rig move, positioning, and pre-loading, the loss is more than $0.5 Million at an oil price of $30/BBL. So the project option that requires less production downtime will always be more attractive.

5.3.3 Selection of the ESP Option

All options were studied and the cost and benefits were estimated for the short and long terms, along with the installation costs.

Using reservoir engineering input, flow streams were generated for five years. Risked increase in oil production was used for increase of revenue (cash inflow), while the estimated expenditure along with the production downtime was considered for the costs (cash outflow). The ESP conversion option had the largest NPV (net present value) and was selected.

A brief description of how NPV is used in capital budgeting and investment planning to analyze the profitability of a projected investment is presented in **Appendix A3**.

There were a few major concerns with the ESP option:

- ESP run life
- Requirement of offshore rig for replacement of failed pump
- Significant production downtime for rig move, positioning, and pre-loading

- Local power generation
- Candidate selection for the ESP recompletion (concerns about abnormal increase in water cut with increase in drawdown)

All these were factored into the economic analysis. In spite of these uncertainties, the economics were very robust for the ESP option. The benefits came from two different ways:

- **Well Benefits**

  It should be possible to produce the wells with more drawdown with ESP than with gas lift. The increased well production is achieved with no additional gas load on the system.

- **System Level Benefits**

  The reduced gas lift requirement would result in increased gas lift supply pressure at the platform. This can be used to improve the lift efficiency of wells by lifting deeper.

  Converting two GL wells to ESP will free up about 6% of the gas lift load of the remote field and about 16% of the load on the platform. The lift gas thus freed up can be used for allocating to other wells in the field or elsewhere, thus making some spare lift gas capacity available for lift optimization of other wells. Freed-up gas can also be used later for development wells.

  This will also reduce the LP gas load on the production system, and therefore the production separator pressure would decrease. All other wells would produce more fluid due to reduced FWHP.

5.3.4 Role of System Optimization Model in Redevelopment

The system optimization model helped to accurately determine the production rates and system pressures for the various options considered, and it played a vital role in studying various redevelopment options. The production rates and pressures could be arrived at only by recognizing the dynamic interaction between the wells, pipelines, and equipment. The operator was already maintaining an updated network system model, so setting up the optimization scenarios was not difficult.

For the option of converting gas lift wells to ESP lift, the production benefits were estimated by well modeling and system modeling, as follows:

- Starting with the calibrated models of gas lifted wells (i.e., using the same inflow), ESP well models were built. Nodal analysis was used to obtain the risked uplift from the individual ESP producers.
- The two gas lifted wells identified for the conversion were replaced with ESP wells in the network model. Results of optimization runs performed using the network model were used to estimate the production increase due to network benefits on all of the platforms of the field.

The network model optimization runs also indicated the resultant production manifold pressure and gas lift header pressure to be expected after the project execution.

The quantity of lift gas freed up by the ESP wells was 7.0 MMscf/Day. As a result, the platform production separator pressure decreased by about 30 psi (from 265 to 235 psig), and the gas lift manifold pressure on the platform improved by 120 psi (from 1080 to 1200 psig).

There were similar, associated benefits from the other three platforms of the field due to reduction of production system pressures and increase of gas lift supply pressure. The system optimization model had predicted an overall increase of about 1600 BOPD from the whole field. This was achieved.

5.3.5 Project Conclusions

1. The ESPs removed part of the bottleneck in the gas lifted production system, causing significant decrease in production system pressure and increase in gas lift supply pressure.
2. The system benefits were more significant than the production benefits from the candidate wells. This allowed further optimization of the field.
3. The well candidate selection played a vital role for the success of this project.
4. Excellent interdisciplinary communication was maintained among the production, reservoir, and workover engineering teams. This played a crucial role in the successful and timely implementation of the project.
5. A decision and risk analysis incorporating the reservoir knowledge and other uncertainties helped to minimize the risks associated with the project.
6. The results from this project confirmed that the overall benefits of any optimization exercise should always be evaluated at the "system" level.

6

Real-time Optimization

With the advent of new digital communication technologies, the industry moved forward and filled the time-gap between the field and the network model. The real-time systems display the status and results of operations, all in one place, to improve ease of access to information and promote transparency. The results from real-time optimization have been very attractive. While computers do the repetitive and massive calculation work, the RTO systems free up valuable human resources to perform more analytical work.

This chapter discusses the following topics:

- Value addition from the digital oilfield
- Need for RTO
- Implementation steps of RTO
- RTO functionalities and architecture
- Well and network management
- Reservoir management
- Calculation and display of asset KPIs
- Daily production monitoring

6.1 Need for Real-time Optimization

6.1.1 Introduction to the Digital Oilfield

The *digital oilfield* is also called "Integrated Operations," "Smart Fields," "Intelligent Oil Field" (*i-Field*), or "Field of the Future."

The concept of synchronizing or integrating various data coming in digital form was originally put into practical use in the eighties by the drilling discipline. The idea was to bring all the live data together to a single place, analyze the data and make prudent decisions then and there, and guide the rig accordingly. Drilling operations, especially offshore and deepwater, are very expensive. Even a few hours of delay or non-productive time of the rig will increase expenses. So, if any change in the drilling program can be decided using the latest available data and the change implemented without any delay, it would save money.

The following people are usually involved in the decision-making process of drilling operations:

- Drilling manager (and sometimes a drilling consultant)
- Reservoir engineer
- Production / completion engineer
- Geologist
- Drilling rig manager
- Other specialists from various service companies supporting the drilling operations

Most of these people (with the exception of rig manager) are in town, and they assemble at the company headquarters onshore for the morning meeting. They make key decisions about the changes to be made in the well program based on data that was gathered during the previous day. The drilling rig awaits this decision before proceeding further.

With the concept of the digital oilfield, an expert opinion from a specialist is immediately available. Therefore, a virtual meeting is possible anytime, and the rig can proceed to the next step, reducing non-productive time significantly. The digital oilfield has provided superior results by connecting all aspects of a drilling operation to ensure productivity, efficiency, and safety. The real-time operations center (drilling) became a big success by the 1980s and became a preferred way of working due to its tangible benefits. RTOC (D) helped companies make drilling decisions using live data analyzed by experts and various stakeholders.

For production discipline, a company that operates offshore platforms producing, processing, and compressing gas has several options on what to do with the gas produced:

- Export some of the produced gas to onshore
- Use some of the gas for reservoir injection
- Circulate some gas for gas lift
- Use some of the gas as compressor fuel if lean fuel gas import is insufficient

The fields being dynamic, there are several variables that can affect the *supply* of gas (well, compressor, and plant availability and processing costs). Other variables can influence its *demand* (contractual commitments, seasonal variations of demand, revenue from excess sale gas, etc.), and these could change every day. A production and export strategy should be decided every morning, based on all the information that controls the supply and demand. If all the up-to-date information is available, the operation strategy can be decided and implemented for the day, honoring the contractual commitments, and maximizing net revenue. A digital oilfield setup is capable of doing the required data processing and projecting the best mode of operation based on a holistic analysis.

Looking at the status of the oil and gas industry at this time, with costs escalating, oil price fluctuating, and skilled manpower shrinking, operations are becoming more complex for exploration and production. The *easy oil* has been mostly produced, and cutting edge technologies are needed to produce the oil left in the ground. Under these circumstances, the preferred approach is to let computers handle tasks like data sifting and quality checking of running equipment and wells and to use the skilled manpower for more creative work. This can be accomplished by setting up real-time optimization that frees up engineering manpower to do more value-adding work such as troubleshooting and problem-solving.

"The Digital Oilfield should be seen as a concept, encompassing not just data being streamed from far distant wells, but the processing, integration and analysis of that data in a number of physical environments. Automation is a vital part of this process. By taking the human out of mundane tasks, productivity is increased, and the expert is freed to use his or her knowledge on more difficult, analytical work. One of the features of the Digital Oilfield is the requirement for collaboration across a wide range of expertise, meaning that decisions are arrived at with a full awareness of all the issues involved, hugely increasing efficiency." [21] Thus, a decision made with a multi-disciplinary approach is now possible almost instantaneously, thanks to the digital oilfield.

Technological innovation has been a prime driver behind the move in the development of the digital oilfield, but it can only work properly when automated operations are streamlined, with work processes embedded with skilled people. A vital component of their success is ensuring that people understand the concept and process, and are well trained and fully committed.

There are several benefits from the digital oilfield. Timely information leads to better decisions and productivity gains, while an increase in the number of remotely operated fields decreases the risk of accidents with fewer people traveling to, and working in, potentially dangerous environments. In addition to this, there are tangible cost savings as well. "Shell believes that the value of its 'Smart Fields' implementation over a five-year period was US$5 billion, while IHS Cera have calculated that by using digital oilfield processes, companies could realize up to 25% savings in operating costs, up to 8% higher production rates, 2–4% lower project costs, and as much as 6% improved resource recovery, all within the first full year of deployment. [21]

The benefits of *i-Field* operations come in the areas of operational efficiency, production optimization, collaboration, decision support, data integration, and workflow automation. The key to success is to set up reliable data communication, customized automation (as needed for the process in the specific field), and a synchronized workforce with sufficient skills and commitment. An information center in the office, equipped with a facility to perform audio/video calls to various production / maintenance locations, is required to initiate discussions and get proper feedback.

As a foundation to the digital oilfield, three important components are required:

- An asset or equipment model
- Live data from the asset or equipment
- A supervisory system with the embedded workflow to interact with the model, review the optimized result, and execute a command in the field based on the optimized output.

The system is usually put in place in the sequence mentioned above.

6.1.2 Difficulties in Continuing with Offline Optimization

Production optimization started as a manual effort performed by engineering and operations every now and then, with the aim of streamlining operations and improving profits. The focus was to improve well performance or correct obvious inefficiencies in some parts of the field.

This manual optimization used field data (well parameters, well test data, equipment status, and facilities data) that was available through production reports from the field in handwritten format or soft copies of spreadsheets. The data could be one or two days old. The engineer-in-charge would review some of the information from the field and initiate optimization in some sections of the operation based on observed anomalies. Due to departmental boundaries, the optimization was not targeted for the full field. The efforts were not repeated on a regular basis due to other operational priorities. No proper record was maintained about the efforts and results of the optimization.

The offline optimization became more systematic with the implementation of a model-based optimization for the entire network. This laid the foundation for a well-structured, all-encompassing, holistic optimization of the entire asset. It brought all the stakeholders together. However, two important aspects of the offline optimization made the process cumbersome and very difficult to continue. These are listed below:

- **Use of human power**:

  This offline process uses valuable time of people at all stages. Several companies find themselves not equipped with staff to be dedicated to continue with this type of optimization. Skilled manpower is expected to be available to address other operational problems (mostly unexpected) arising in the field.

- **Process cycle time**:

  If the field has hundreds of wells and large facilities, one full optimization cycle can take several weeks. Field experience indicates that, for a large field, even with a fully committed team, one cycle of complete offline optimization would take about three months. But the well production rates and water cut can change significantly in three months. By the time the model is updated, the model might not be "current" in some aspects. A few things would have changed in the field by then.

Continuing the system optimization in the offline mode would be very challenging. This would get pushed back due to other priorities in field operations. A faster and more accurate optimization with more automated workflows is required. The solution is *online optimization.*

6.1.3 Moving into Online Optimization

If the optimization workflow is automated, two obvious benefits can be achieved:

- Less of skilled people's time will be used up for field optimization
- The optimization cycle can be repeated more often.

All six steps of the optimization process mentioned in *Section 4.4.5 System Optimization Workflow* can be performed, but most of them would be done by a computer-based software, instead of being done manually. This helps reduce the amount of skilled manpower involved in the process and the time taken for a cycle. With more frequent optimization, the gap between the field and the model would be reduced. The model would more or less represent the field at all times. A sustained level of optimization would be maintained, as shown in **Fig. 6-1**.

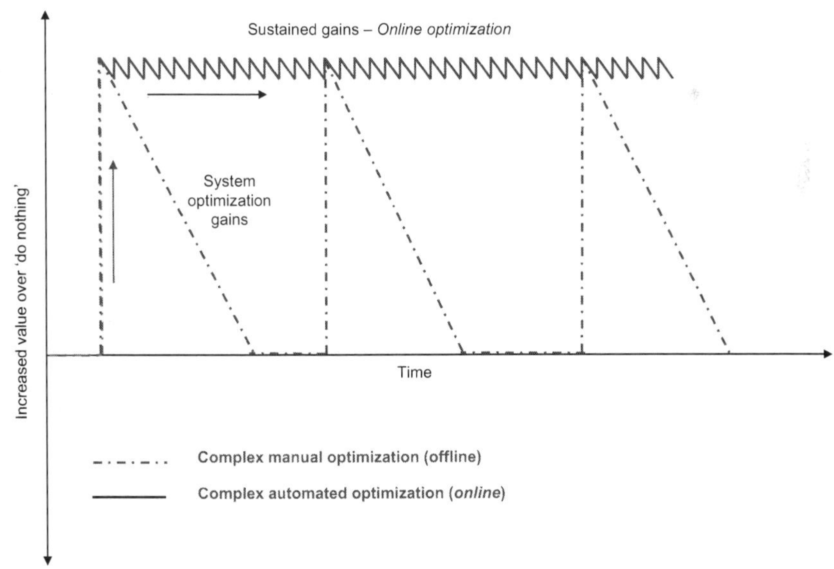

Fig. 6–1. Sustaining gains with automated optimization

The gains from optimization tend to revert to norm as the system changes. Therefore, automation of the process is key to *sustain* the gains. This provides the necessary justification for the automation of the workflow. Note that the gains shown in the plot in **Fig. 6-1** should be read in a qualitative manner. The gains from initial optimization could be substantial, but the gains from subsequent optimization cycles may not be that huge. However, there are always sufficient gains that can justify the effort that goes into each optimization cycle.

It is not the cost of the real-time supervisory system itself, but the customization required while setting up the online optimization workflow that will determine the benefit from the system. The intelligent real-time optimization software will have some common features that are applicable to all fields. In addition to this, the modeling consultants would configure asset-specific workflows, developed with input from the engineering and operations staff of the operating company. Every field has its own unique workflows and business processes. The online optimization workflow should address and satisfy the unique needs of the field.

A simplified, core workflow of a typical RTO is shown in **Fig. 6-2**.

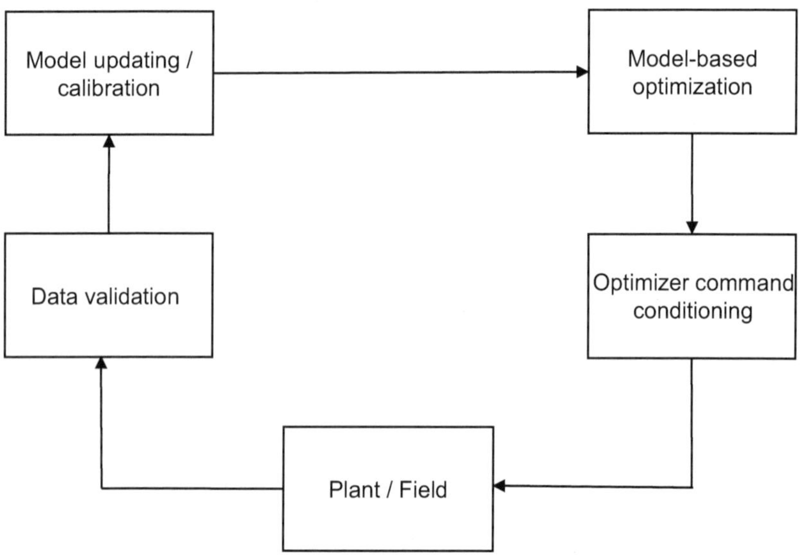

Fig. 6–2. Process workflow of a RTO System

The optimization team would be able to download the online model (well and network models) at any time, either to perform investigations in offline mode or make corrections in the model and place it back online.

6.1.4 Benefits of Intelligent Real-time Optimization System

Apart from the human effort, the real-time system requires some investment in instrumentation and information technology setup (hardware) and the visualization and optimization software. As the organization embraces the new way of working, the quality of operations and efficiency would improve. Apart from the benefits over manual optimization, the presence of digital data and powerful tools handling the data would open up new avenues of improvement in the operation of the asset with new technology.

The following are the broad rewards for this work. The first three are tangible benefits. The last one is intangible, but that is sometimes more inspiring than the first three items.

1) **Increased Production**:

   As the field conditions change, the real-time systems perform optimization more frequently. The production gains from optimization are sustained by increased frequency of optimization. With a daily optimization approach, the asset is operated at near-optimal conditions at all times.

2) **Improved Uptime (Reduced Downtime)**:

   Avoidance or reduction of downtime of wells and equipment translates into more revenue. The system can detect sub-optimal performance of a well or piece of equipment at the preliminary stage itself. This makes it possible to address the issue before it could cause big downtime. Production problems that took 3–4 days to rectify in the traditional way of working are getting corrected now in 1–2 days. The well intervention required to correct the problem may still be same, but the time taken for diagnosis, remedial work procedures, and mobilization has been significantly reduced. This translates to avoiding more than 50% of well downtime. This is achievable as the people behind all these activities are interacting on a real-time basis. Furthermore, it is often easy to correct a problem at its initial stage.

3) **Reduced Operating Costs**:

   With more preventive maintenance and real-time surveillance, the unexpected shutdown of equipment is minimized or avoided, if possible. With the daily production strategy being decided based on the actual availability of asset components, the concept of maximizing net profit is almost always achieved. This reduces the costs to a minimum possible level. Moreover, delivering production commitments in terms of volume and quality helps retain a brand name and reputation.

4) **Improved Operational Efficiency**:

   The following features improve the operational efficiency of the asset:

   - The system helps people from different departments work together as a team. The improved team culture breaks down departmental barriers within the organization. By working together, they maintain good communication. This leads to improved operational efficiency.

- ○ A well-informed, multi-disciplinary team manages the asset in a proactive manner. They are able to handle process upsets with minimal downtime.
- ○ The RTO system reduces the number of field visits required for people, thus reducing human exposure to the hazards of a harsh working environment.
- ○ Engineering time will be saved for high-value work, as the routine tasks of surveillance and data verification are now performed by the intelligent systems.
- ○ The system is usually set up as a web application so that the asset information is available to members or experts located anywhere in the world.
- ○ The real-time system makes a single set of integrated information available to all. All people see *the same data at the same time.* All the information required by the company's asset management process is available at a single location. This becomes the foundation for a collaborative decision-making process.

A direct comparison between the offline and online workflows may be summarized as follows:

Offline Workflow	Online Workflow
Process takes months	Process takes minutes
Skilled resources used for managing the entire workflow process	Workflow fully automated (skilled resources may be used, if required for any intervention or validation of results)

6.2 RTO Implementation

6.2.1 RTO System Selection

It is necessary to select the right software that can handle the real-time optimization of the asset efficiently:

- It should be technically competent, versatile, comprehensive, stable, scalable, expandable for the future, and customizable for changing real-time operation requirements. It should have provisions to add additional features that may be required for the company's operations in future.
- It should be easily integrated with the existing IT systems in the company.
- It should comply with the data security protocols of the operating company.

- The most important aspect is that it should be easy to use. An average user should be able to access, review, and extract data or results easily in the form of pre-defined graphs and reports. The user should also be able to generate customized reports instantaneously.

It may be necessary to plan an official visit to another company where the service company has already installed an RTO system that is working. This gives the operating company first-hand information of how the system is working in the other organization and how the operating company is gaining value out of it. It will be necessary to overcome the issues of data security and confidentiality in arranging such a visit.

6.2.2 RTO Implementation Steps

The following steps should be considered when implementing a real-time optimization system:

1. Once the software is selected and the contract signed off, the functional design specifications (FDS) of the proposed system should be finalized. The contract should allow further refinement of the FDS, if needed before the closure of the project.
2. Field and office visits should be planned for the consultants to meet with the operations and engineering teams. The consultants should become fully knowledgeable in the offline optimization process already being used by the operator. They should also visit the location of existing data servers and data communication systems.
3. The new servers to be used for real-time data handling, surveillance, and optimization should be procured after their specifications are finalized and agreed upon.
4. The service company usually works on the RTO configuration in offline mode to build the system and test it independently. For new product development, a factory acceptance test (FAT) may be needed.
5. Once most of the configuration work is done, real-time data is loaded to the software in a parallel mode (the software is receiving data and performing its functions, but results are not communicated to the field). The functioning of the system is observed.
6. Once all the workflows are performed by the parallel RTO in an accurate and satisfactory manner and the system is stable, the RTO system is brought online.
7. The system should be run under observation for a prescribed period (usually for one or two weeks), being continuously monitored on a 24x7 basis by the service company personnel. All functionalities are thoroughly checked out during this time.
8. A site acceptance test (SAT) and user acceptance test (UAT) are performed. The system performance should match with or exceed

the requirements defined in the FDS. All results and observations are documented, and the system is handed over to the operating company. Any functional shortcoming should be captured, highlighted, and rectified.

9. An RTO user training session should be conducted for all designated users.
10. Licenses are configured and access to the RTO system is given to users, as required by the operating company, with different levels of privileges.
    - *Basic User*: Can view all result and configuration screens
    - *Power User*: Has access to all modules, can update well and equipment models
    - *System Admin*: Can change configuration, add or delete models

After going online, a DESC (Design and Evaluation Services for Clients) engineer may be put in place in the operating company office to provide any expert support for 3-6 months or one year, until all users become well-versed in the use of the system. This is optional and would be decided as per the contract.

The company should encourage all individuals to use the new online system to make the best use of it and use this as a tool of communication between different teams. This would encourage all staff members—both traditional and innovative thinkers—to use real-time optimization systems.

As the user base increases, there will be findings on how to reconfigure workflows in a smarter or more productive way or how to add new workflows. These would be short-listed, prioritized, and implemented as required by the business.

6.2.3 RTO Implementation Work Schedule

The implementation of a real-time optimization system should be considered when the operator is conversant with the use of model-based network optimization and has streamlined operations around the offline optimization process.

To move on to the next natural step of online optimization, apart from the calibrated network model, there are two additional requirements to make the RTO functional:

- First, a data-gathering system that should be put in place to collect and transmit real-time data from the field.
- Second, a smart supervisory system with customized workflows that should be installed to oversee the online optimization process.

These two activities would be required as a part of the implementation of the RTO system.

The execution plan for a typical online system is shown in **Fig. 6-3**. The plan shows how closely the operating company and the optimization consultants should work throughout the implementation.

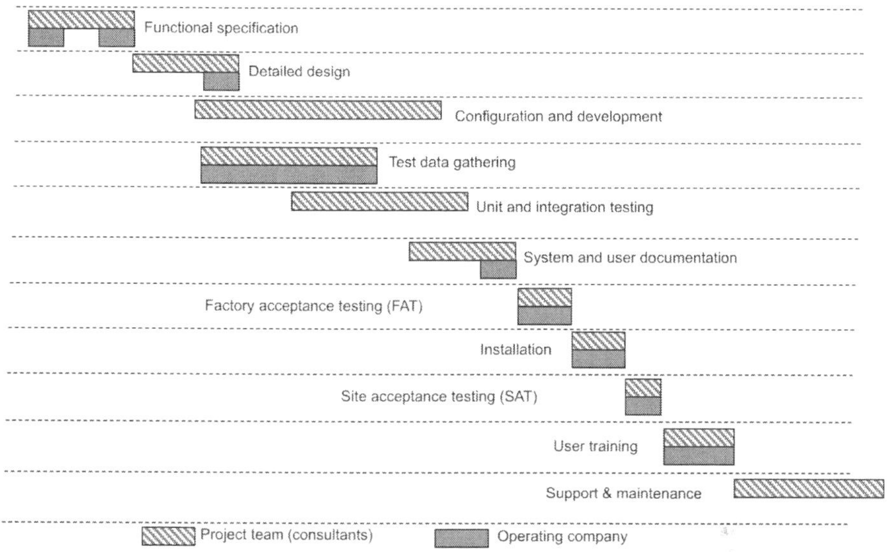

Fig. 6–3. Execution plan—online optimization system

After establishing offline optimization, many fields have implemented real-time data surveillance and real-time optimization, either separately or together. This has been well documented. [(9), (10), (11)]

6.2.4 Functional Design Specifications

A document describing the functional design specifications of the intelligent RTO system is prepared jointly by the operating and service company teams. This should describe the full set of workflows to be performed by the real-time optimization system, covering all the activities and deliverables. Some of these work flows may be default features of the software, and the rest may be configured specifically for the asset in focus. The Engineering and Operations team members should review this document and ensure that it is comprehensive enough for the present and future business requirements of the asset.

All the desirable workflows that are considered valuable by the operating company should be included, even if a few of them may not be used immediately. Some of these may be required to achieve the future vision of the company. This could also include additional reports, trend charts, displays, etc., as envisioned by the company to achieve more transparency and efficiency in operations.

The RTO consultants assign a unique *tag number* for every field data point that feeds data to the real-time system. The data may be used for a direct digital display in a trend plot or for backend calculations. Details of the configuration work should be documented properly for future reference.

The RTO system configuration is validated by the team that authored the FDS document. The team reviews the functioning of the system and confirms that the system delivered meets the requirements spelled out in the FDS.

6.2.5 Getting Real-time Data

Getting real-time data from some areas of the field is tough. This is more applicable to old fields, as some of these fields were designed for manual surveillance. In old fields, companies initiate a *data to desktop* (D2D) project for the purpose of bringing digital data from field nodes to the desktop in real time. In new fields, the hardware for data collection and transmission is already in place, or provisions have been made for the same.

In general, most of the critical pieces of equipment are located at the Facilities Platform (offshore) or Facilities Cluster (onshore). The data from critical equipment and separators is routed to the control room, where the computers are usually located. The system controlling separation, compression, or the gas treatment process may be of proprietary or old design. A copy of the live data can be tapped off the system, to be used for real-time visualization or production optimization.

But the challenge remains for wells and field pipelines. At desert locations, a well pad has a Christmas tree with a simple control kit consisting of a control panel, hydraulic system, and battery charger. There is a valve near the fence in the departing pipeline. The production from the well flows in buried pipelines. People drive to this location and note the wellhead parameters manually in log sheets.

Running a cable to bring the data from hundreds of such wells to the Central Facilities Control room may be very difficult. Sandy or hilly terrain may pose severe challenges to the extent that running cables may not be possible in some terrains. In such locations, wireless methods may be used to transfer the data. Offshore, usually some sort of data communication infrastructure is already in place for communication between satellite and facilities platforms and from offshore to the onshore base. This may be used.

Data collection and transmission for wells and field pipelines is critical. The operating company should make this one-time investment by initiating a field survey and accomplishing the data transmission using appropriate technology. Otherwise, optimization will not be complete.

In the olden days, the data transmitted was very limited (a few morning reports, spreadsheets, and emails). For real-time systems, a large quantity of data transmission will be required back and forth. The bandwidth of the communication should be designed or expanded to send and receive large pockets of real-time data and commands.

The amount of data transmitted in real time should be decided based on business requirements. A transmitter may send pressure data once per minute or once every second. The former option sends 1440 data points a day, whereas the latter option sends 86,400 data points a day from one sensor. So if 10,000 data transmitters are actively sending pressure, temperature, and flow rate data, the resulting volume of real-time data will be massive. The communication bandwidth should be set up accordingly. It is a good practice to have a larger capacity available, so that the system can handle future requirements as well.

A video conferencing facility is needed for the asset team members to meet virtually from different locations. It is useful to get direct feedback from everyone (especially field operations or maintenance) when a problem is encountered in the field. This helps establish a way forward, after taking input from all the parties using a multi-disciplinary team approach, which is the foundation for success in system optimization.

6.2.6 Frequency of Optimization

Even though frequent optimization helps keep the field in the best operating condition, there is a time interval required for the wells and the network to stabilize after a change is made. This is a function of the size of the network and the number of wells connected to the network. In large networks with hundreds of wells, if a change is made in one part of the network, it takes more time for the effect to cascade to the full network.

The pre-set routine daily optimization may be performed either at 6:00 AM or midnight. This is useful for formulating and implementing production and injection strategies for the day.

- For pressures and temperatures at critical nodes, the averages of the values over the last 24 hours would be used.
- The production and injection volumes measured during the previous 24 hours (cumulative volumes from totalizers) may be used for comparison with model results.

Non-routine optimization can be performed at any time by downloading the model and performing the run.

6.2.7 Well On/Off Criterion

A method should be developed for the real-time system to decide if a well is on or off. This is required to estimate the well uptime. This also should be automated. Using the well status and uptime, the volume of production or injection during the day is calculated. If the well produces at a rate of 2000 STB/Day and the uptime is 18 hours, the fluid produced would be assigned as 1500 STB for the day.

In producer and injector wells, normally there is at least one valve that is connected to the emergency shutdown (ESD) system. This would probably be a wing valve, actuated by a pneumatic, hydraulic, or electrical actuator. If this valve is shut, there is no flow from the well. However, if the valve is open, it does not mean that the well is fully on, as a choke may be used to control the flow or bring the well up to full flow in steps. So the status of the valve combined with a flow rate is considered for the system to calculate the uptime of the well. Sometimes it is more complicated because the well may not reach its full flow rate when the normal choke opening is reached. This should also be factored in when deciding well uptime. It should be set up based on a logic that is suitable for the field.

The following logic may be used to compute well uptime in gas lifted fields with real-time optimization:

- Gas lifted producer wells:
  - Tubing and casing wing valves open
  - Well injecting at least 75% of the set point gas lift injection rate
  - (Gas lift manifold pressure-casing pressure) > 50 psi
- Water injection wells
  - Wing valve open
  - Rate > 50% of the normal injection rate

The water injection rate is measured continuously by turbine meters installed in the water injection lines of the wells. Instantaneous data on water injection flow rates is available. Therefore, the system knows the rates once the well starts injecting. The cumulative 24-hour injection data comes from the totalizers.

6.2.8 Acceptance Tests

The acceptance test consists of two components: User acceptance test (UAT) and Site acceptance test (SAT). The former is signed off by the user after witnessing the functionalities being performed by the real-time system; the latter after site implementation of the system and handover—that is, after running the system "live" for a specified number of days.

The precise details of the tests are elaborated in the acceptance test in line with the FDS. It will be ticked off item by item. At a minimum, the tests would cover the following areas:

- Basic functionality tests—Access to the system and user configuration are established as required.
- System configuration—Operating company data is correctly configured in the system.
- Display—The system display of the asset at various levels is functioning as required by the FDS.
- System integration—The system properly communicates with other software products of the operating company.
- Engineering tests—The asset well models are correctly tuned in accordance with the recalibration methods described in the FDS document.
- Error/Exception handling—Error conditions are correctly handled.

6.2.9 Phased Implementation

The key to success in a real-time optimization implementation lies in handing over the accountability of the system to the staff of the operating company. The operating company can get the long-term benefits of the system only if their staff use all the features of the system. To achieve this goal, some companies prefer a phased implementation performed in two stages: Stage 1 for wells and Stage 2 for the network. Stage 1 is rolled out first, and users are given access to the live system. While the consultants are configuring Stage 2, the users become familiar with the system installed in Stage 1. They would then be ready to move to Stage 2 once it is rolled out.

6.2.10 RTO for Intelligent Wells

The generic term "intelligent well" is used to indicate the presence of an extra arrangement within the well completion that gives the capability of performing two additional functions that differentiate an intelligent well from a conventional well:

- It is capable of collecting, transmitting, and analyzing wellbore production and reservoir and completion integrity data
- It may also allow remote action to control reservoir, well, and production processes

At first, technology became available to monitor wells with downhole sensors. This gave real-time information about what was happening in the well. This laid the foundation for "automating" the well. The reliability of these monitoring systems has improved over the years. Then came the ability to control the downhole production.

Complex horizontal and multi-lateral wells in multiple-zone reservoirs are nowadays completed as intelligent wells. Downhole pressure and temperature sensors and multiphase flowmeters connected to the surface through optical cables give the capability of monitoring the downhole production parameters remotely. The inflow control devices (ICD) / downhole chokes and hydraulic control lines in the completion allow the remote capability of controlling production. By operating inflow control devices, production from various reservoir zones could be altered to achieve the desired liquid rate or water cut. By changing the artificial lift settings (e.g., ESP speed), the overall production can be adjusted.

The key objectives of the intelligent well systems are summarized below:

- Improved recovery (optimize for zonal/manifold pressures, water cuts, and sweep)
- Improved zonal/areal recovery monitoring and allocation (locate remaining oil and define infill development targets)
- Optimized production (achieve improved lift, acceleration, and reduced project life)

- Minimized capital investment to exploit an asset
- Reduced intervention and operating costs
- Optimized water handling

Intelligent-well technology delivers improved hydrocarbon production and reserves recovery with fewer wells. It also improves the efficiency of water-floods and gas-floods in heterogeneous or multilayered reservoirs when applied to injection wells, production wells, or both. The production and reservoir data acquired with downhole sensors improves the understanding of reservoir behavior and assists in the appropriate selection of infill drilling locations and well designs. Intelligent-well technology can enable a single well to do the job of several wells, whether through controlled commingling of zones, monitoring and control of multiple laterals, or even allowing the well to take on multiple simultaneous functions: injection well, observation well, and production well. [22]

With such instrumented wells with controllable devices, the production strategy becomes more complicated. But this can be included as part of the workflow in real-time system optimization.

If a zone is closed off partially in a smart well using ICD, the inflow changes. If the well is gas lifted, the lift requirement for the well will also change. The real-time system updates the well model with the change of inflow. Optimization runs determine the new optimal gas lift requirement for the well. The system model adjusts the gas lift supply to the referenced well. On a real-time cycle, the system can optimize itself within a reasonable period. Thus, the RTO system can perform production optimization for gas lifted intelligent wells more efficiently.

6.3 Functions of Intelligent Systems

At a minimum, the intelligent system that supports real-time operations should perform several broad functions falling under the following categories:

- Well engineering
- Well management
- Network management
- Reservoir management

6.3.1 Management by Exception

The real-time optimization system usually follows *Management by exception* (MBE)—a style of business management that focuses on identifying and handling cases that deviate from the norm. Thus, one priority task of the system is to identify the components that have deviated from the expected functional performance.

If there are 50 wells under one engineer's area of responsibility, and if two wells have abnormally low flowline temperatures, the system will highlight those two wells on the screen. This way, what is abnormal would catch the attention of the engineer first. The engineer would discuss with the field operations team and check why the temperatures are lower than normal values, before reviewing data from the rest of the wells in the assigned area. It could be due to low production from the well or faulty instrumentation.

The same approach is applied throughout the system to catch inefficiency as quickly as possible. The aim is to identify and correct the problem with minimal downtime.

6.3.2 Well Engineering—Well Model Calibration

The system receives production well test data and calibrates well models automatically.

Production well tests are validated based on a pre-set criterion. If the validation is successful, the test is used to calibrate the well model.

The model calibration method is determined based on the type of well and the nature of the reservoir. In water-flooded reservoirs, the primary method of calibration may be based on the change in productivity (or injectivity) index of the well, keeping the reservoir pressure unchanged.

For a naturally flowing well, when a new well test is obtained, the model performs a top-down analysis (i.e., pressure drop analysis, starting with the flowing wellhead pressure) and determines P_{wf}. The multi-phase flow correlation selected for the well uses the measured flow rate, fluid properties, and tubing size for the hydraulics calculation. For this reason, the selected correlation should be validated with a recent flowing gradient survey to confirm its suitability for the well.

The model then runs an operating point analysis to confirm the operating point (production rate and FBHP) and adjusts the PI *within limits* to match the operating point with the P_{wf} calculated in the top-down analysis. If the change in PI is beyond acceptable limits, the calibration will fail. The production engineer should retune the model manually and reload the tuned model back into the online system. The engineer may initiate additional downhole surveys for the well, if required.

For a gas lift well, the calculation is a bit more complicated. Using the surface casing pressure, the gas lift pressure at depth of gas lift injection is calculated in the casing-tubing annulus. Simultaneously, a top-down pressure drop calculation is performed for the tubing side to obtain the tubing pressure traverse. Thus, the model verifies the positive pressure differential required for gas lift injection at the designed gas lift depth, as shown in **Fig. 6-4**. The gas injection rate is calculated using upstream and downstream pressures and the orifice size at the designed operating depth in the well. This would be compared with the measured gas lift rate reported in the well test.

If the gas lift injection rate so calculated does not match with the actual injection rate, the orifice size or the surface gas lift injection pressure may be taken as a variable and sensitivity runs performed. If orifice size is varied, the orifice port size that gives a reasonable match for the lift gas rate would be estimated and compared against the actual port size installed in the well. If the calculated port size is within the calibration limits, the model recalibration will be completed successfully. But if it is outside the limits, the well model calibration will be abandoned and a report generated. The orifice may be either plugged up or eroded—or the measured gas lift rate or casing pressure might need to be counterchecked.

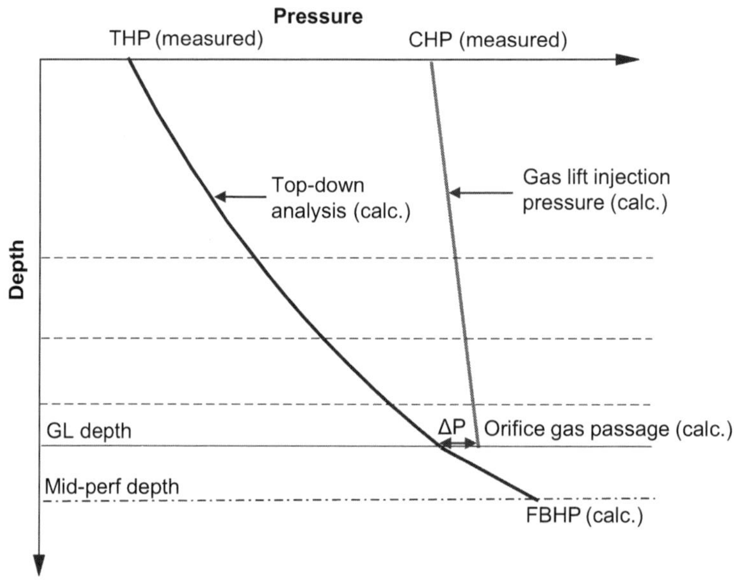

Fig. 6–4. Gas lifted well model successful calibration

If the differential pressure existing at the designed operating depth is either very small (insufficient for gas injection) or there is none, it is an indication of gas lift not working as designed. In **Fig. 6-5**, the tubing pressure at depth is actually more than the casing pressure, indicating that the well may be lifting from one or more shallow mandrel(s). The RTO system will abandon the well model calibration in this case and generate an error report.

Once the well model is recalibrated with a new well test, the calibrated well model will replace the old model in the real-time server. A new well performance curve is generated and attached to the respective wellhead object. This ensures that the new performance curves are used for all calculations, both at the well and network levels. The old well model will be archived.

Water cut values fed from the real-time data system may also update the well model and generate a new performance curve.

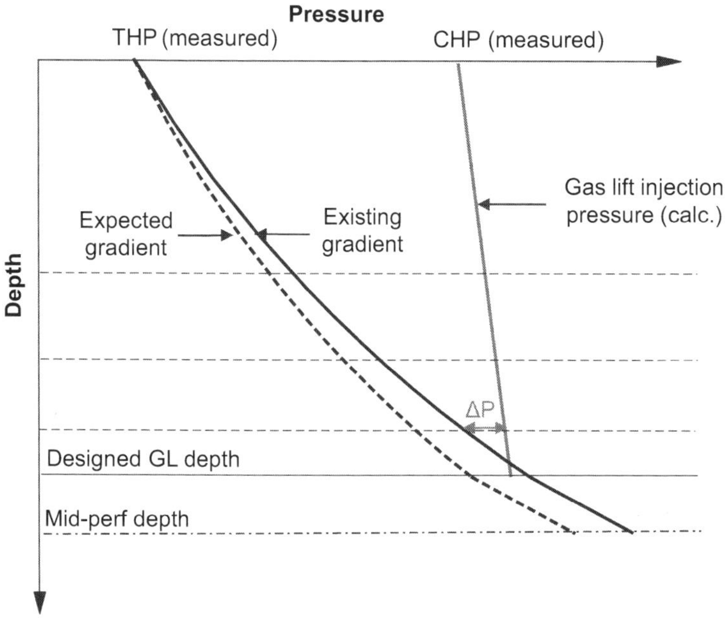

Fig. 6–5. Model calibration failure due to shallow or multi-point gas lift injection

6.3.3 Well Management

6.3.3.1 Daily Rate Predictions

Using the wellhead pressure, gas lift injection rate, and casing head pressure, averaged over 24 hours, the real-time system calculates the produced volumes of oil, gas, and water for the number of hours the well was "on" during the last 24 hours. The latest well model will be used for this calculation. The real-time system can also display daily or hourly average values or spot values.

6.3.3.2 Monitoring Well Performance

The estimated volumes of production or injection during the last 24 hours are compared against measured values, in the case of wells equipped with continuous flow measurement. These would be compared against the well capacity determined by the recent well test, if the volumes are not measured continuously. Should the well volumes fall short of the estimated volumes, alert messages are sent out. These wells can be ranked by severity. Wells with significant anomalies would attract immediate attention. This function is available for both producers and injectors.

6.3.3.3 Trend Plotting for Analysis

Graphs can be generated by plotting key parameters used for well and surface network monitoring. Other than the pre-defined formats, it is possible to configure additional templates or generate customized graphs or reports from the real-time data, as and when needed. A rich set of controls for configuring and exporting charts is available. The system is versatile in examining and comparing the data of multiple wells on the same plot, if required.

6.3.3.4 Production Allocation

The quantity of oil produced, as claimed by the field, often does not match with the oil received at the storage area. This is due to errors in well test measurement, uptime calculation, and shrinkage factors. This should be reconciled. In traditional reservoir management, actual production metered in the field at the sales point as STB is taken as final, and not the production claimed by production operations. An RTO system can allocate field production volumes back to the wells.

Consider a 6-well Cluster C with a production capacity of 10,000 STB Oil/Day. One well in this cluster is down for four hours during the day. Using the *uptime*, the system calculates the predicted oil production of the cluster as 9500 STB.

The volume of oil produced and delivered by Cluster C to the storage terminal during the past 24 hours is accurately known from meters. This data is also fed to the RTO system. If the actual quantity of oil received at the storage terminal is only 9150 STB, the system will back allocate the actual oil production of 9150 STB to all six wells, as shown in **Table 6-1**.

Table 6–1. Production allocation

Well number	Oil production capacity, STB Oil/Day	Well uptime, hours	Estimated production during last 24 hours, STB Oil	Production back allocated, STB Oil	Allocation factor
C01	1000	24	1000	963	
C02	1650	24	1650	1589	
C03	1350	24	1350	1300	
C04	3000	20	2500	2408	
C05	1200	24	1200	1156	
C06	1800	24	1800	1734	
Total	10,000		9500	9150	0.9632

The system allocates the production on a pro-rata basis and comes up with an allocation factor for every well or group of wells (0.9632 in this case). As can be seen, there is an approximation applied here during the back allocation. In reality, the shortfall (350 STB of oil) from cluster C may be from any of the wells. But there is no way of knowing this as the wells are not tested every day. No abnormality was detected by the surveillance system, and no other downtime was reported. Taking the uptime for the well C04 as 20 hours, the production shortage has been distributed among all wells in the cluster.

6.3.3.5 Target Monitoring and Alerts

Another useful function of the RTO system is target monitoring for the wells in a lease. This can be accomplished for production or injection, either on a well-by-well, cluster, or field basis. With input values for the *allowables* and *technical rates* for each well, the system can help identify under- or overperforming wells (**Fig. 6-6**). This gives an idea of the source of shortfall, if any, and helps engineers focus on those wells that have fallen short of target rates. List of wells not meeting target rates can also be obtained in a tabular form and ranked based on severity. Once identified, engineering and operations teams would work towards improving the production or injection in those wells.

- *Allowables* refer to the production or injection rate recommended by the reservoir surveillance team. These rates have upper and lower limit relaxations.
- The *technical rate* of a well is defined as the maximum allowable production rate, as prescribed by the reservoir management team. Producing the wells at rates beyond technical rates may affect the effective hydrocarbon recovery from the reservoir. The technical rates are reviewed by the reservoir team on a regular basis.

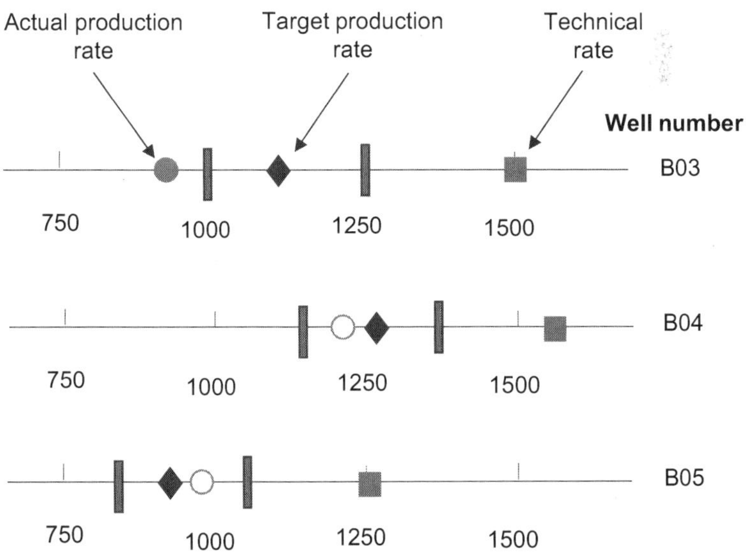

Fig. 6–6. Target monitoring display

In this example, production from wells B04 and B05 are closer to the target rates. The actual production rate of well B03 is lower than the lower target limit, hence the circle got highlighted.

As the average and total production can be calculated on a well-by-well, field, or asset basis, the shortfall or excess is automatically available at any point in time, and the production can be adjusted in the field accordingly. This helps ensure enough oil is in storage before the tanker arrives.

6.3.4 Network Management

6.3.4.1 Network Engineering

The changes made in the field for production lineup, change of well status, equipment in service, etc. get updated in the network automatically. The RTO system performs optimization runs daily, using these updated asset models. The field constraints already input into the model are honored. The model sends revised set points as output to achieve production and injection targets. Typical results include:

- Choke settings for NF wells
- Gas lift rates for GL wells
- Power frequency for ESP wells
- Optimum speed for compressors
- Production separator pressure

Similar optimization runs can be performed for managing any ad-hoc operation. When planning for a compressor shutdown or a pipeline pigging operation, optimization runs can be made using the updated network model, and the best optimized operation for the scenario can be selected. The result includes new set points for each scenario. It also is possible to compare results from such scenarios versus actuals for a selection of optimum set points. The system would back up all network models for various scenarios, downloadable for further offline analysis, if required.

6.3.4.2 Field Factor

A network model predicts production rate from the total field. The overall production predicted for the previous day could be compared with the actual production metered in the field. The actual production may be lower than the estimated production. The field factor is calculated as follows:

$$\text{Field Factor} = \frac{\textit{Actual production volume}}{\textit{Estimated production volume}}$$

A field factor in the range of 0.95–1.0 is considered good. If the field factor is low, then the model prediction is much higher than actual production. Some possible reasons are:

- Water cut changes at well level are not checked regularly or reported properly; water cut measurement method may be inadequate.
- Production well test measurements are not accurate enough or well test frequency is insufficient.
- Model calibration accuracy is low (multiphase flow correlation selection or other settings in the model require review).
- Some wells are producing less, but the surveillance system has not captured it.
- Well downtime has not been reported properly.

6.3.5 Reservoir Management

One of the reservoir management functions the real-time system can do is voidage control. Voidage replacement refers to replacing the volume of oil, gas, and water produced from the reservoir by injected fluids.

The voidage replacement ratio (VRR) is defined as the ratio of reservoir barrels of injected fluid to reservoir barrels of produced fluid. "VRR is fundamental for comprehending the injection and production balance as it profoundly affects the pressure distribution within the reservoir and evidently the wells production rate." [23]

VRR is also known as fluid-in fluid-out (FIFO) or injection-withdrawal ratio (IWR).

$$\text{Voidage replacement ratio} = \frac{Volume\,of\ fluids\ injected\ (reservoir\ volume)}{Volume\,of\ fluids\ produced\ (reservoir\ volume)}$$

Reservoir volumes for each production well for each day will be calculated as follows:

$$Vp = NpBo + GpBg + WpBw - NpRsBg$$

Where:

Np, Gp, Wp: production stock tank volumes (oil, water, gas)

Bo,Bg,Bw: formation volume factors (oil, gas, water)

Rs: solution GOR

Reservoir volumes for the injection wells can be calculated as:

$$Vi = GiBg + WiBw$$

Where:

Gi, Wi: injection standard or stock tank volumes (gas, water)

Bg,Bw: formation volume factors (gas, water)

The online system uses the PVT data of the fluid model from the applicable well model for each well. The reservoir pressure and temperature are picked up from the well model and used for the volume calculation.

Fig. 6-7 shows the VRR calculated for a day, using average rates. The daily flow rates are either estimated from a physics-based model or predicted by a model based on machine learning (ML). Usually, VRR calculation is done for a longer period (three or six months). This will give a true picture of overall production and injection status in the field, as single-day rates may be affected by well downtime during the day. The VRR can be displayed on a daily basis or for a period of time.

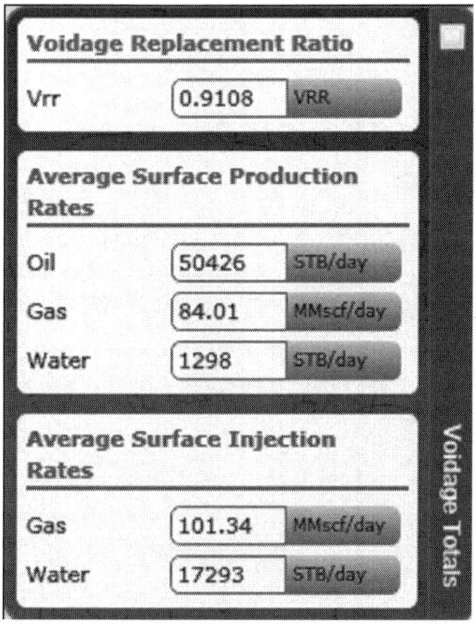

Fig. 6–7. VRR from daily rates

Of the many types of injection patterns used in reservoirs, the peripheral and 5-spot injection patterns are more common (**Fig. 6-8**).

Typical VRR values required for these injection patterns are as follows:

Peripheral pattern : 1. 1–1.4

5-Spot pattern : 1.0–1.1

If the reservoir is huge and has varied properties, it may be hypothetically divided into compartments for the purpose of voidage management. The VRR can be calculated for each compartment. If one injector is supporting more than one compartment, its contribution can be split between the compartments. This can be configured while setting up the model with input from the reservoir engineering team.

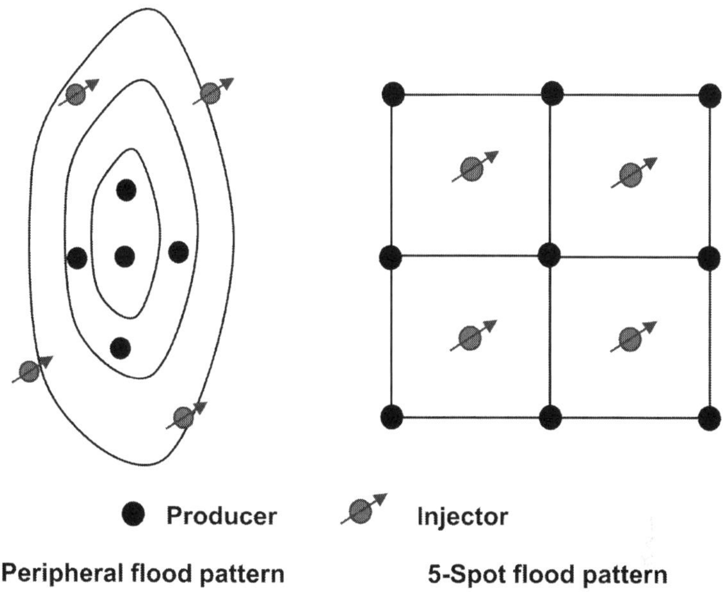

Fig. 6–8. Reservoir injection patterns

A background reservoir map with well locations is used for the display of VRR monitoring. For easy identification, the well locations are marked on the base map. Different icons are used to differentiate types of wells (producer, gas injector, water injector, and WAG injector).

The VRR management is a powerful surveillance tool for maintaining reservoir pressures in a healthy manner, thus improving the ultimate recovery of hydrocarbons.

6.4 Tracking KPIs using RTO

A key performance indicator (KPI) is a measurable value that demonstrates how effectively a company is achieving key business objectives. KPIs provide a focus for strategic and operational improvement and create an analytical basis for decision-making. They help the team members stay focused on what matters most. Organizations use KPIs to evaluate their success at reaching targets. Managing assets with the use of KPIs includes setting targets (the desired level of performance) and tracking progress against that target.

Once the key business metrics have been identified and agreed at a team level, the team would want to track them using a real-time reporting tool. KPI management can be done using a dashboard reporting software, giving the entire organization insights into the team's current performance. This can also be provided by the intelligent supervisory system that handles the real-time optimization and manages the dataflow, or by any other software that will pull and integrate data from the RTO system and display the results.

The biggest benefit is the transparency of the whole system and the avoidance of sudden surprises to anyone. The transparency is helpful as everyone contributes to achieve the target. The results being displayed on a common platform takes out the guesswork and gives no room for individual interpretation or manipulation.

6.4.1 Field-specific KPIs

A few KPIs taken from an oil and gas asset that implemented real-time optimization are discussed in detail.

The asset exports the oil and compresses and injects the produced gas into the reservoir. The asset also injects water to augment reservoir pressure support. The field uses water injector wells, gas injector wells, and WAG (water alternating gas) wells for reservoir pressure maintenance.

For the asset-based organization, a total of six KPIs have been established. These are monitored on a daily basis through the RTO system.

1. Maintain well production rates within the agreed quarterly allowables, without exceeding the well technical rate for more than seven days.
2. Achieve VRR of 1.2 (+10%) for Asset A and 1.0 (+10%) for Asset B while adhering to other reservoir management guidelines. Values in parentheses indicate *stretch targets*, normally included to challenge and motivate people.
3. Maintain >95% rotating equipment availability by predicting early failures through trend analysis and preventive maintenance (excluding planned maintenance shutdowns).
4. Achieve a minimum of 90% of the annual well testing plan (test once every quarter at a minimum per well) while maintaining <20% (stretch <10%) well test rejection.
5. Achieve 100% annulus "A" remote pressure monitoring for the total existing 94 oil producer strings per month and maintain MAASP wells below 10% per month. MAASP refers to the maximum allowable annulus surface pressure.
6. Minimize production loss from wells, clusters, and system upsets (from healthy, non-problematic wells) by maximizing wells' availability remotely and restore well availability within seven hours.

6.4.2 KPI on Production Target

The first three KPIs are illustrated. The first KPI is about achieving production targets:

Main key objective	Field KPI
Production optimization	Maintain well production rates within the agreed / issued quarterly allowables without exceeding the well technical rate for more than seven days.

If the field had less well availability (i.e., more downtime) than planned during a month, it will be necessary to increase production from a few "booster" wells to achieve the field production targets. If the field gets into this situation, the KPI cautions that operations may produce a few wells at rates exceeding allowable rates, but that too cannot be done for more than seven days. At any point in time, the technical rates cannot be exceeded.

6.4.3 KPI on Voidage Replacement

The RTO system has the data of volume of fluids produced and injected during the last 24 hours. By converting the surface volumes to reservoir conditions, it computes the daily VRR and also the cumulative VRR and displays them on screens for the team to understand and catch up. It also stores the values in its database.

Main key objective	Field KPI
Improvement in reservoir monitoring	Achieve VRR of 1.2 (+10%) for Asset A and 1.0 (+10%) for Asset B while adhering to reservoir management guidelines

The average VRR achieved by Asset A in about the first 100 days of the year is shown in **Fig. 6-9**. As can be seen, the VRR has improved gradually, but has reached only 1.15, whereas the target is to achieve or exceed 1.20. Some additional attention is required to improve the VRR.

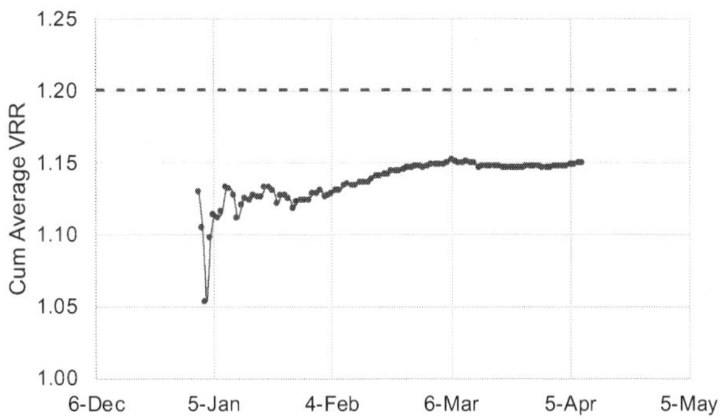

Fig. 6–9. Cumulative average VRR

6.4.4 KPI on Equipment Availability

The third KPI is about equipment reliability and availability. The goal is to maintain the availability at more than 95%.

Main key objective	Field KPI
Improving the equipment reliability	Maintain >95% rotating equipment availability by predicting early failures through trend analysis (excluding planned maintenance shutdowns)

In all these cases, the RTO system will calculate the KPIs using the data from the past 24 hours and monthly or yearly cumulative values up to the previous day. The engineering and operations staff, the technical support team, and the management will see the dashboard displays, which will show the results for the KPI under each category. This will inform them whether they are in line with the targets or about to miss the targets. This helps improve their focus in those areas.

6.5 Architecture of RTO

6.5.1 Dataflow Overview

Fig. 6-10 shows a basic real-time optimization setup. The data collected from all the instrumented parts of the asset (wells, pipelines, equipment, plants, etc.) is routed to a local server, and this data gets transmitted to the real-time server (usually located at the main office). This server feeds the data to an intelligent supervisory system. Many automated workflows that are pre-programmed are executed by the supervisory system. The field KPIs that contain the field targets are also fed to the supervisory system. The supervisory system processes all this information and displays the results.

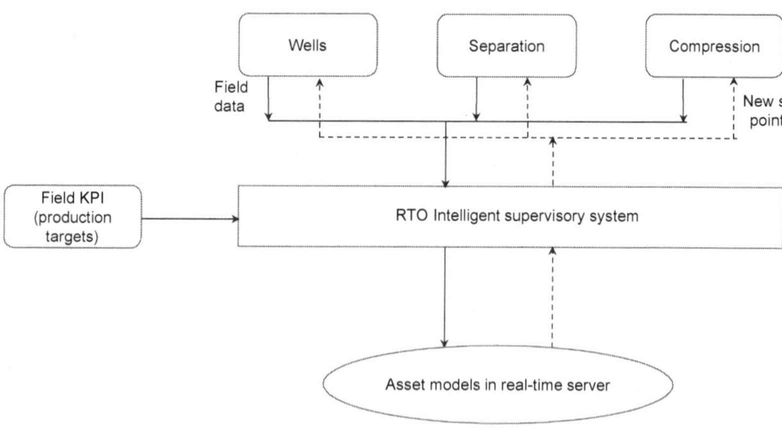

Fig. 6–10. Simple RTO dataflow

The supervisory system triggers optimization runs of the asset models. The model results are fed back to the supervisory system. Depending on the type of results, an automated / semi-automated / manual quality check may be performed on the results before implementing the new set points in the field. The quality check could be done in the office or in the field.

6.5.2 Functional Capabilities

The intelligent daily optimization system should have the following capabilities:

1. The supervisory system should be intelligent enough to distinguish between real and spurious data from the data stream flowing from the field. If the data is bad due to voltage fluctuations or problems related to electronic instrumentation, the faulty data should be ignored.
2. The real-time system should send a warning message to the engineer-in-charge if the asset model does not converge or if the results output by the model are outside the acceptable range and hence rejected by the system. While the system would continue working with the old set points, it would invite attention from the engineer or field staff responsible for the operations.
3. The system should be capable of keeping a record of the history of the old and new set points. It will also have sufficient memory to store the field data for a predefined duration of time. This will allow the user to check the data trend for, say, the last one month or so.
4. The system should be capable of displaying the field data of selected parameters, along with the set points in parentheses. All engineering and operations staff who are authorized will be able to see the actual field conditions against the set points with regards to wells and equipment. For those parameters that deviate significantly from the set point values, message alerts would be displayed, highlighting the abnormality.
5. The system may also communicate a log of activities to the engineer concerned at regular intervals so that the engineer is kept informed of the changes that have been suggested.
6. Based on the company's requirements, some of the live field data may be displayed in monitors at various locations / terminals in the main office and field offices. This shows the status of the field operations. Visitors can also see this data. This is mostly to boost the image of the company.

6.5.3 RTO Schematic

The deployment of the online optimization system is to achieve data connectivity and functional performance, as shown in **Fig. 6-11.** [3]

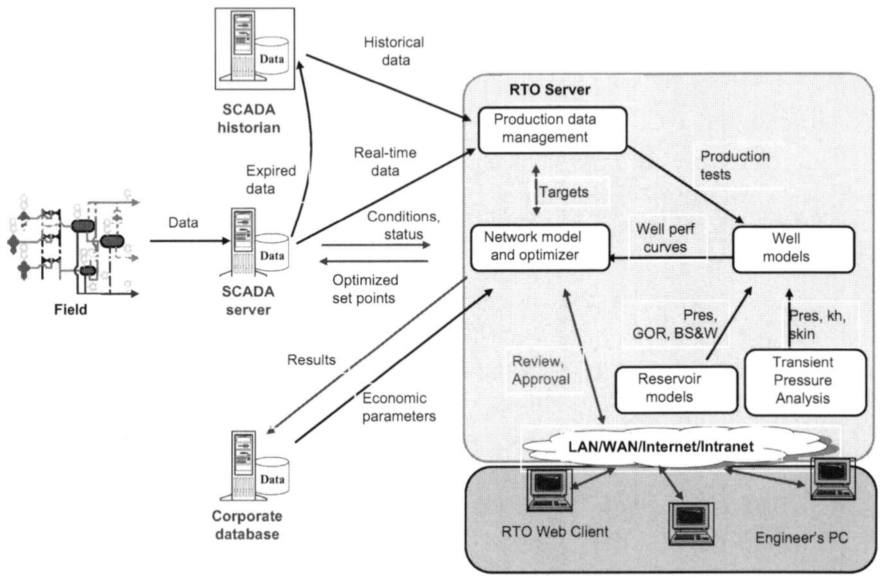

Fig. 6–11. Connectivity of RTO with data systems and servers

Real-time data is being sent from the field to the SCADA server.

- The RTO Intelligent server receives real-time data from the SCADA server and historical data from the SCADA historian. It stores all data, interfaces with external applications, and automates the work flow.
- It passes on quality-checked new well test data to the well models and sends a signal to recalibrate the models. It monitors which models got successfully recalibrated and which models failed in the calibration process and sends out success / failure warning messages.
- It creates performance curves for the well models calibrated successfully.
- It gives signals to run optimization runs with network models, either as a pre-set routine or due to an operational requirement.
- It ensures that the results (new set points) are communicated to the field via the configured validation system in place.

If the model recommendations settings get implemented directly, the RTO is said to be in a closed loop.

6.5.4 RTO System Architecture

The RTO software solution should be designed to seamlessly integrate with enterprise IT architecture. One way of accomplishing this is by conforming to the following principles during the development of the system. Other methods may also be followed.

- Open Software Standards—open and easily integrated technology should be preferred at all places in the system where communication is required with enterprise systems. For example, non-proprietary universal standards such as XML and XML Web Services may be employed to communicate at the enterprise level over HTTP.
- Simple Deployment—the RTO user interface is built using server-generated web pages. Consequently, system deployment is trivial, and the system is easily customized to conform to client user interface standards.
- Component-Based Design—software is developed using industry standard component technology, with components conforming to standard interfaces, allowing software components to be replaced by enterprise components that fulfill the same function.

Using this approach, the high-level system structure may be designed as shown in **Fig. 6-12**.

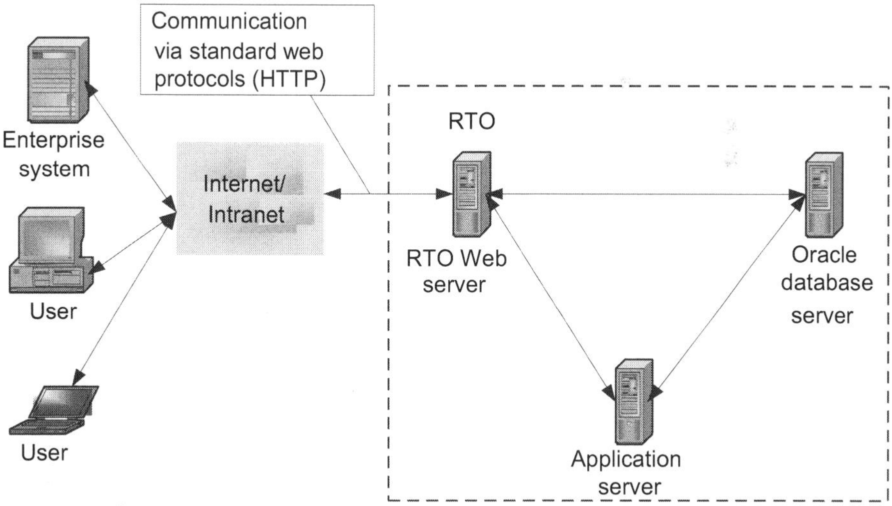

Fig. 6–12. RTO system overview

The RTO System shown above consists of three logical parts:

- RTO Web Server—processes the RTO user interface and handles requests via web-based protocols for real-time optimization functionality

- Application Server—powers the engineering applications that are included within the solution (for example, network model) and the system functions for accessing and managing these applications
- Database Server—powers the oracle database that RTO uses to maintain the system state and the functionality required to access and manage that data

Note that this is a logical view of the system deployment model and does not necessarily reflect the actual hardware architecture required. The system is designed such that each of these components may scale to meet heavier system demand. Alternatively, some or all of the servers may be collocated where this is required.

The installations done now use the latest technology available today and are more robust in terms of connectivity and performance. The IT specialists will advise as to the best options to be used in any project.

6.6 Data Visualization

6.6.1 Types of Information to Display

The visualization tools are very effective in communicating a message to the viewers instantaneously, and in a very convincing manner. This is usually offered by the service provider of the RTO system. A plot is created with live data for an engineering investigation and subsequent discussion. A graphing tool can be used to display the trend of a selected parameter or a group of parameters. XY graphs, bubble maps, bar graphs, pie charts, or other formats may be used to highlight the emphasis of the results.

Even though the results may be obtained through complex calculations in the background, it is very important to ensure that the final result presentation is simple and straightforward so that any viewer can grasp the message quickly. To this effect, the display should be kept simple and easy to follow.

Real-time systems may display:

- Measured data
- Calculated data
- Asset management data
- Business results update

Information such as volume of sales, production / injection targets, equipment status, pipeline status, and HSE performance tracking can be displayed.

The display content (what to display) and format (how to display) are selected based on the business objectives of the organization.

Some sample visualizations follow.

6.6.2 Asset Performance at a Glance

This is the first page of the display showing the high-level summary of everything that is happening now or that happened in the field in the past 24 hours. Many people look at this screen in the morning and take away different messages based on their interest and focus.

- The production operations team focuses on the shortfall of production and injection volumes of the previous day and the reasons for the same (e.g., wells that remained shut-in or equipment that remained shutdown). They think about actions they can initiate to make up for the loss in production / injection.
- The production engineers focus on the wells that are not performing at full potential and possible remedial work that can be done to get the wells back to full-capacity production or injection.
- The maintenance engineer looks at process components that are still under maintenance and reviews what can be done to make it available as soon as possible.
- The team leaders look at the asset performance as a whole as well as the KPI status that day and in the context of the month and year.

This sheet should be designed to show all the key performance parameters at a quick glance. *Keeping it simple* is the key. The network overview display screens from two different assets, A and B, are given in **Fig. 6-13** and **Fig. 6-14**, respectively. The equipment status is also included in the colored display.

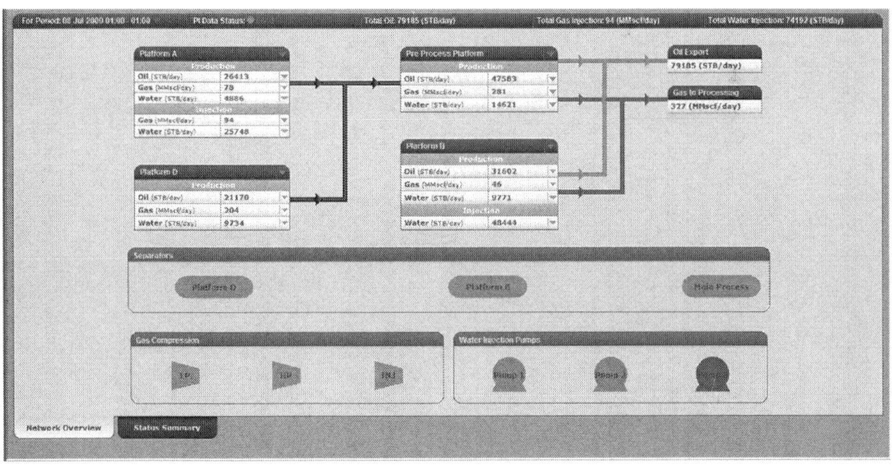

Fig. 6–13. Network overview (Asset A)

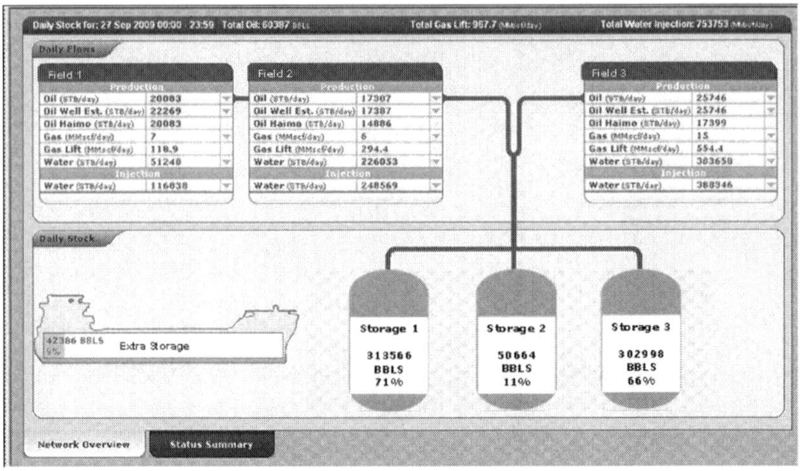

Fig. 6–14. Network overview (Asset B)

Those who are interested in knowing more details on any particular item may click on the icon and drill down to view additional information.

6.6.3 KPI Presentation

The updated status of KPIs can be viewed by clicking the KPI status option. The screen shot shown in **Fig. 6-15** is not perfect, as it tries to present too much information in a single screen. Ideally, less information on a page results in more effective communication.

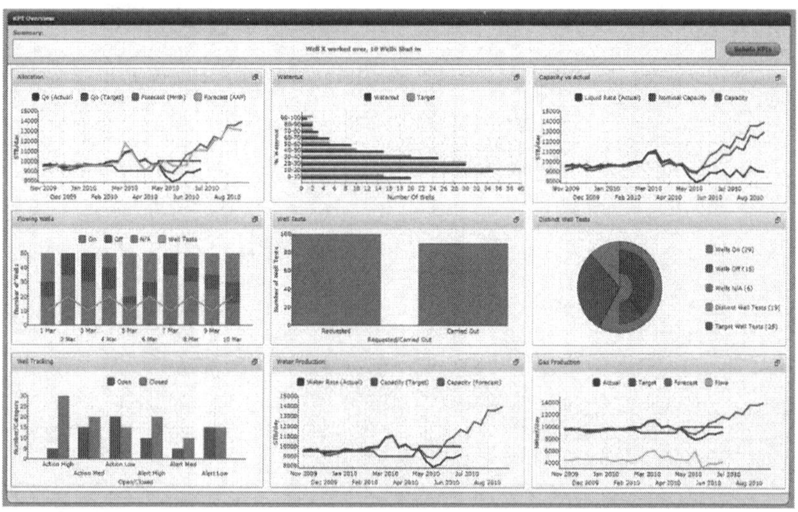

Fig. 6–15. KPI status

A simple display of oil production status is shown in **Fig. 6-16**. While the graph shows the actual vs. target vs. forecast oil production, the bar charts show the difference in the daily rates and the cumulative oil production.

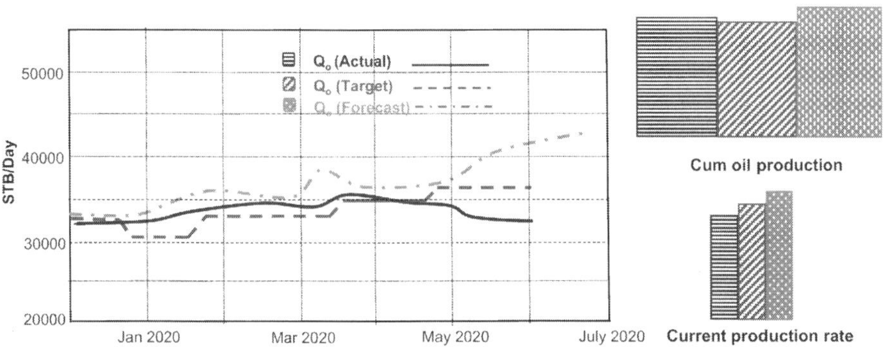

Fig. 6–16. Oil production status update

- The production curves trace the trend of production from the beginning of the year.
- The bar charts indicate that the cumulative oil production so far is more than the target, but is short of the forecast. The current daily production rate indicates that the actual production rate has gone below the target rate.

6.6.4 Well Test Validation and Well Model Tuning

The screen in **Fig. 6-17** shows the number of well tests received in the past 24 hours and how many of them failed in the validation process. The failed ones are highlighted to catch attention.

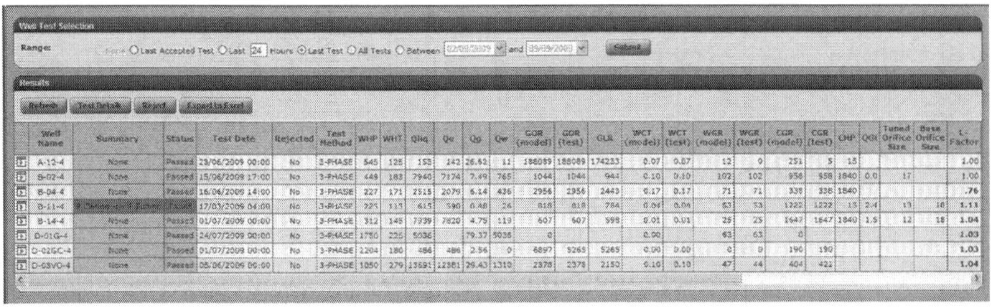

Well Name	Summary	Status	Test Date	Rejected	Test Method	WHP	WHT	Qliq	Qo	Qg	Qw	GOR (model)	GOR (test)	GLR	WCT (model)	WCT (test)	WGR (model)	WGR (test)	CGR (model)	CGR (test)	CHP	QGI	Tuned Orifice Size	Base Orifice Size	L-Factor
A-12-4	None	Passed	23/06/2009 00:00	No	3-PHASE	545	125	153	142	26.62	11	188089	188089	174233	0.07	0.07	12	0	251	5	15				1.00
B-02-4	None	Passed	15/06/2009 17:00	No	3-PHASE	449	183	7940	7174	7.49	765	1044	1044	944	0.10	0.10	102	102	958	958	1840	0.0	17		1.00
B-04-4	None	Passed	16/06/2009 14:00	No	3-PHASE	227	171	2515	2079	6.14	436	2956	2956	2443	0.17	0.17	71	71	338	338	1840				.76
B-11-4	[illegible]	Failed	17/03/2009 04:00	No	3-PHASE	225	115	615	590	0.48	26	818	818	784	0.04	0.04	53	53	1222	1222	15	2.4	13	18	1.11
B-14-4	None	Passed	01/07/2009 00:00	No	3-PHASE	312	145	7939	7820	4.75	119	607	607	598	0.01	0.01	25	25	1647	1647	1840	1.5	12	18	1.04
D-01G-4	None	Passed	24/07/2009 00:00	No	3-PHASE	1750	225	5036		79.37	5036	0			0.00		63	63	0						1.03
D-02GC-4	None	Passed	01/07/2009 00:00	No	3-PHASE	2204	180	486	486	2.56	0	6897	5265	5265	0.00	0.00	0	0	190	190					1.03
D-03VO-4	None	Passed	05/06/2009 00:00	No	3-PHASE	1050	279	13691	12381	29.43	1310	2378	2378	2150	0.10	0.10	47	44	404	421					1.04

Fig. 6–17. Automatic well test validation results

6.6.5 Production Well Test Data Trend Plot

The well test data can be viewed in a tabular format as well as in the form of a plot. The variables to be included in the plot can be pre-formatted or selected as required. This gives an opportunity to visually recognize the relation between changes, e.g., how the other well responded when the gas lift injection rate was changed, and so on. The P_{wf} calculated by the model while calibrating the model using the test can also be included in the plot (**Fig. 6-18**). The P_{wf} measured using a well test with a flowing gradient survey can also be shown with a special symbol so as to infuse confidence in the P_{wf} values calculated by the model, before and after.

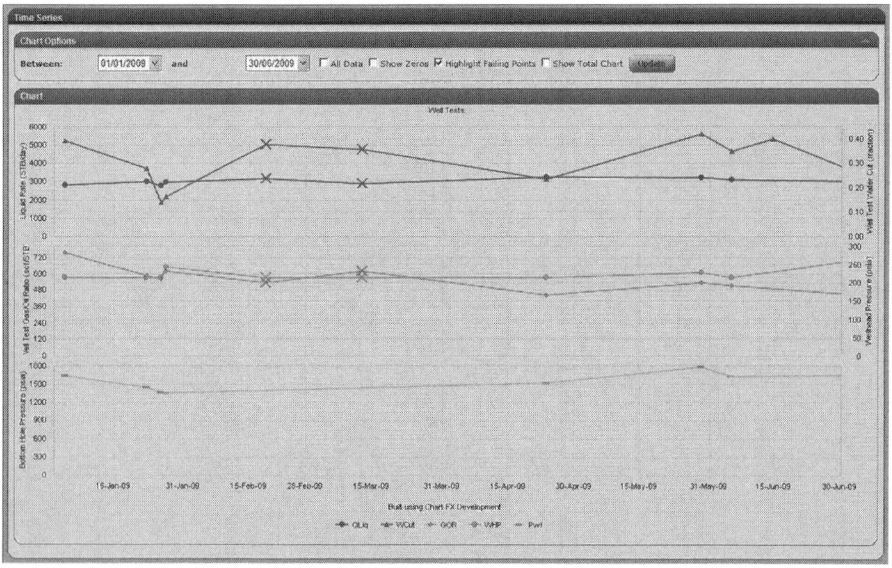

Fig. 6–18. Trend plot of well test with flowing bottom-hole pressure

6.6.6 Daily Production Monitoring

The wells with flowing parameters showing significant deviation when compared to the calculated parameters are flagged for attention (**Fig. 6-19**).

Visualization plots can be configured to review the performance of pipelines and various pieces of equipment in the model. The production and injection data can be analyzed extensively for individual or a group of wells almost instantaneously.

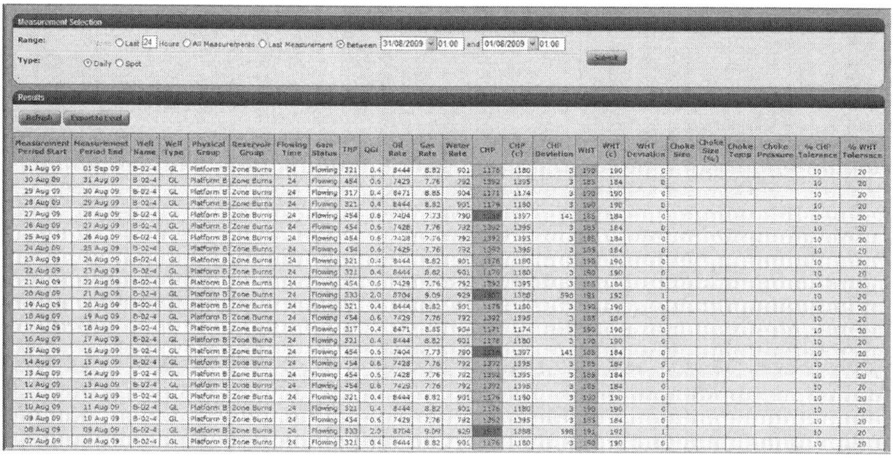

Fig. 6–19. Production monitoring

7

Optimization of Unconventional and Subsea Gas Lift Wells

The most recent challenge encountered by the gas lift industry is the optimization of unconventional and deepwater wells. The wells in unconventional pays require a different treatment with regard to their production and artificial lift strategies. With two or three lift methods being considered for the life cycle of the well, there are unique challenges in the design, operation, and optimization of gas lift in unconventional wells. Gas lift for subsea wells, on the other hand, requires a high pressure downhole outfit, with barrier-type isolation. Deepwater gas lift installations mainly use single-point gas lift and riser-base gas lift, but have problems of flow assurance. Understanding the specifics of these gas lift applications will help improve production optimization of these assets.

This chapter discusses the following topics:

- Well design and artificial lift consideration in unconventional pays
- Production optimization of unconventional wells
- Improved gas lift designs for unconventional wells
- Single-point gas lift system in subsea wells
- Deepwater gas lift well equipment
- Riser base gas lift

7.1 Optimization of Unconventional GL Wells

Gas lift operation of unconventional wells and deepwater wells requires a different treatment from conventional wells.

- The nature of well construction, completion, hydraulic fracturing treatment, and the decline pattern of wells in tight, unconventional reservoirs requires a very different consideration to determine the artificial lift requirement of these wells, compared to traditional onshore wells.
- The flow assurance issues associated with deepwater operations and the cost of well intervention in this environment make the gas lift application of subsea wells very different from typical offshore gas lift wells.

Application of gas lift to these two types of wells is briefly discussed in this section.

7.1.1 Production Strategy for Unconventional Wells

Unconventional reservoirs have low porosity and permeability. The production rates have been observed to drop by over 40 to 80% within the first year in the unconventional reservoir wells. These are long horizontal wells drilled in very tight reservoirs that require a massive hydraulic fracturing ("frac") treatment to initiate production. Technology improvements have allowed the industry to continue drilling these wells for oil and gas production. When providing artificial lift to these unconventional wells, conventional gas lift knowledge cannot be applied to these wells.

Optimal well designs vary by area and operator in unconventional plays, but drilling and completion trends generally favor longer lateral sections, more hydraulic fracturing stages, and greater frac densities per stage. However, while the large contact areas and high reservoir conductivities achieved with multistage fracturing technology dramatically increase initial production rates, they also exacerbate decline rates from horizontal wells. (24)

To give a practical example of the decline, based on information from more than 5,800 wells, it has been reported that at five years of age, horizontal wells in the Permian see decline rates fall to an average of 16.7% and produce just under or about 30 BPD. (25)

The typical decline pattern of a well from an unconventional reservoir is shown in **Fig. 7-1**.

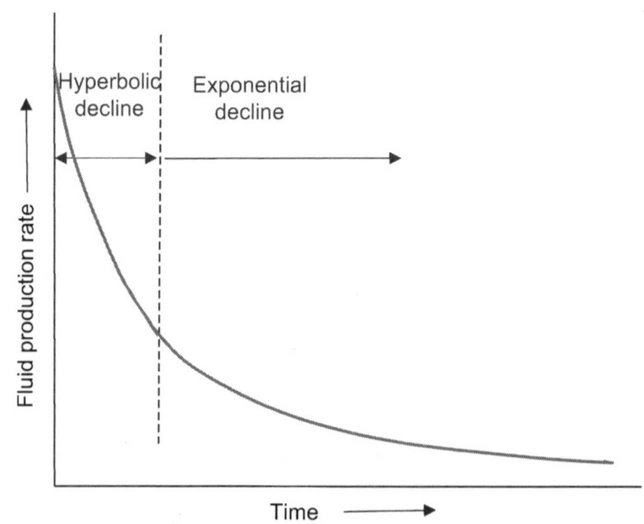

Fig. 7–1. Decline of unconventional wells

Three distinct stages of production decline may be observed in unconventional wells:

- **Stage 1**

  The well after a massive frac treatment starts off with huge production rate on natural flow. Some operators use jet pumps to recover most of the proppants pumped during the frac job. The jet pump is needed only if the proppants could not be recovered with the natural flow from the well after treatment.

- **Stage 2**

  With production on, the reservoir pressure declines, and the PI also declines. The well reaches a stage where it will really be benefitted by artificial lift. If artificial lift is not installed at this stage, the well may cease to flow. Usually, a large-volume artificial lift (gas lift, ESP, or jet pump) may be installed at this stage.

- **Stage 3**

  With continuing decline in production, the well reaches a stage where the type of artificial lift should be changed because the rates have dropped. This time, depending on well conditions and fluid properties, a low-volume lift method such as a beam pump or a progressive cavity pump may be installed; continuous gas lift may be converted to intermittent gas lift or plunger lift. Depending on the well configuration, capillary foam systems or gas injection may be performed to sweep liquid through laterals. Gas injection is mostly used for de-liquefaction of shale gas wells.

Statistics from 2018 indicate the percentages of unconventional wells that used various artificial lift systems are as follows: (26)

Gas Lift	: 40%
ESP	: 36%
Rod lift	: 13%
Plunger lift	: 7%
Jet pumps	: 4%

While operations differ from field to field, some operators follow a random sequence of operations:

- Flow back
- Put the well on ESP
- Remove the ESP and put the well on gas lift
- Remove the gas lift and put the well on rod pump

The operation becomes expensive and shortens the life of the well. If sufficient attention is not paid to streamlining the surface facilities and flowlines, as much as 25% of the productivity may be lost due to high back pressure. This is especially true during the initial high production period of these wells. If a well is allowed to flow into a system of 150 psia at surface, it will have gas expansion that works as natural lift for the well. This leads to a significant increase in production, compared to flowing against higher flowing wellhead pressures. Proper planning is required to avoid excessive back pressure in flowlines.

Fig. 7-2 displays the results of nodal analysis gas lift modeling for a well in the second stage of decline. Even at this stage, decreasing the flowing wellhead pressure from 300 to 150 psia increases the fluid production rate by 25%. A decrease of 150 psi in FWHP causes a decrease of 335 psi in FBHP in this well. The loss of production would be very severe during the first phase of decline.

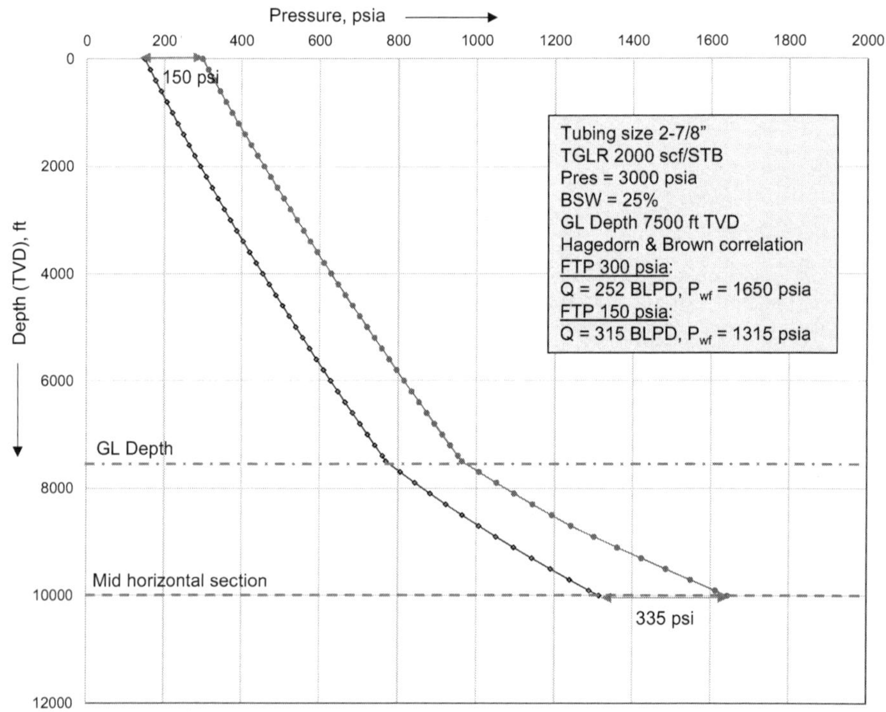

Fig. 7–2. Effect of back pressure

Chemical injection can be done down the casing annulus if gas lift is installed. If gas lift is the chosen method based on well characteristics and fluid properties, a strategy making use of a "total well approach" may be considered optimal for the life of the well, as highlighted below: (27)

- Install one completion that will last for the life of the well
- High-rate gas lift for initial production with artificial lift
- Mid-rate gas lift
- High-rate plunger-assisted gas lift (500 BPD)

- Plunger lift
- Low-rate gas assisted plunger lift (200 BPD)
- Staged plunger lift

For formations with a very low GOR, gas lift may not be the best choice as the plunger lift option for the later part of the life will not work. However, the underlying message is that a proper artificial lift strategy should be considered for the entire life of the well.

7.1.2 Well Design and AL Considerations in Unconventional Pays

7.1.2.1 Well Planning

One of the key lessons learned in unconventional oil production is that the entire well program, including well planning, drilling, completion, fracture treatment, production, and abandonment, should be integrated with the expected production profile for the entire life of the well. The type of artificial lift that would be used should be treated as a part of the well program. It is preferable to change the artificial lift without pulling the completion out of hole. This means the original completion equipment should be planned with the required flexibility.

7.1.2.2 Multi-phase Flow in Horizontal Section

The three commonly encountered wellbore trajectories mentioned below are shown in **Fig. 7-3**.

In a *toe-up well,* the lateral true vertical depth (TVD) gain is negative or the TVD at the toe is less than the TVD at the heel. The downward inclination of the lateral from the toe back toward the build section allows liquids to drain into the heel section easily. In other words, the single liquid accumulation point is closer to the heel.

In a *toe-down well,* the TVD at the toe is more than the TVD at the heel. The single liquid accumulation point would be farthest from the kickoff in this case.

In a *hybrid well,* there are several liquid accumulation points along the lateral. Flow in the horizontal section is difficult in case of toe-down and hybrid wells, especially if the inflow is weak.

Fig. 7-4 shows typical multiphase flow behavior in an undulating horizontal well. In the downward flow sections, the gas/liquid flows are stratified normally. In a stratified flow regime, the fluids are separated into different layers, with lighter fluids flowing above heavier fluids. A stratified flow is more efficient and desirable. This improves the ease of production.

In the upward sections, at low- to medium-fluid velocities, slug flow is the common flow pattern because of liquid accumulation caused by gravity. Slug flow has alternating flow of gas pockets and liquid slugs and shows resistance to flow.

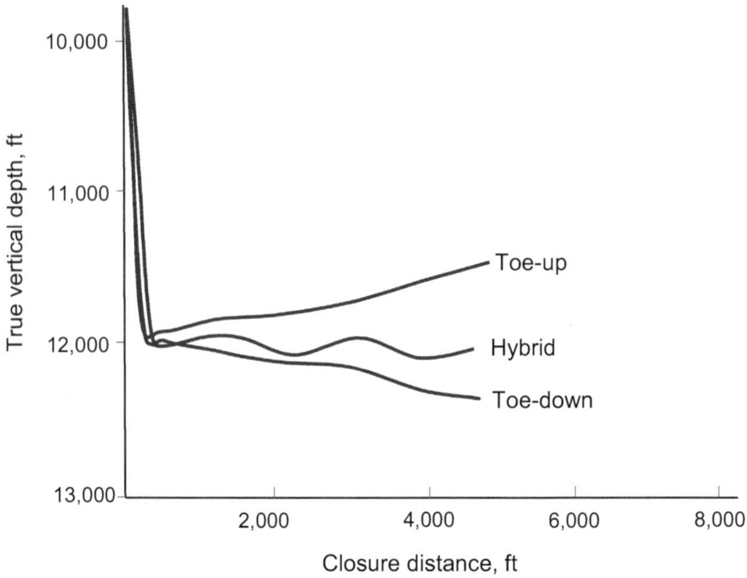

Fig. 7–3. Wellbore geometry affecting well performance

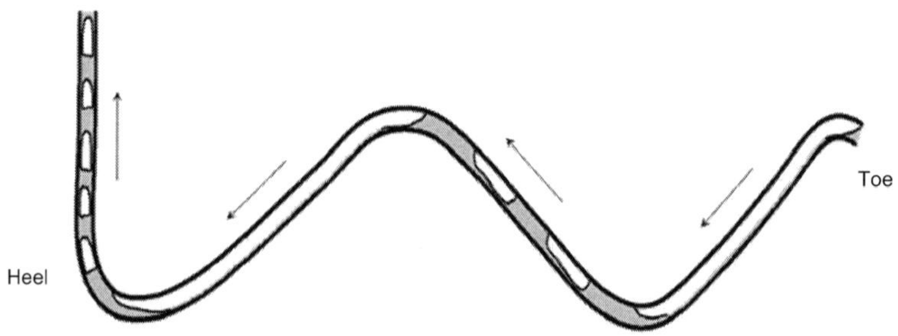

Fig. 7–4. Multiphase flow in undulated horizontal sections

In the horizontal section, 1° toe-up results in stratified flow even with low superficial velocities of the liquid phase, whereas a 1° toe-down leads to slug flow. [28]

Results of the fluid flow experiments conducted in the Tulsa University Horizontal Well Artificial Lift Project (TUHWALP) confirmed that a toe-up horizontal hole is far more efficient in draining liquids from the horizontal section than a toe-down horizontal hole.

Considering the multiphase flow behavior and pressure gradient difference in the upward and downward sections, a toe-up horizontal well profile is preferred. As the toe-up hole facilitates smooth production, it is important to build this into the drilling plan of the well. However, due to several reasons, it may not be possible to achieve the ideal wellbore trajectory in some wells.

7.1.2.3 Position of Lift Depth

The artificial lift equipment can be installed in the vertical or deviated / tangent section above the heel of the well.

The configuration of unconventional wells usually has three sections.

- The vertical section of the well
- A slant buildup section with appropriate dog leg severity (DLS) to turn the well horizontal
- The horizontal section about 4000–6000 ft. long (or more) that will be hydraulically fractured

Because of the rapid production decline in unconventional horizontal wells, the artificial lift system selected must be flexible enough to operate over a wide range of production rates, or the system should be easily resized or replaced. Gas lift is preferred, especially in wells with relatively high GORs. Casing-head instability probably will be more severe with horizontal well gas lift because of the significant gas volume stored in the lateral section. The gas lift valve should be installed above the heel section.

7.1.2.4 Selection of the First AL Method

ESPs have been helpful in getting large production rates from unconventional wells at the initial phase of artificial lift, but there have been many failures due to large quantities of proppants being produced and ESPs operating at high temperatures. Solids and insufficient motor cooling are two main factors for the ESP failures. For this reason, many operators deploy continuous flow gas lift as the first lift method in unconventional wells and later change to intermittent lift—plunger lift or gas lift assisted plunger lift.

7.1.2.5 Selection of Casing Size

For unconventional wells, many operators use 2-7/8″ tubing, in order to produce larger volumes of fluid during the initial production period, instead of 2-3/8″ tubing. If 2-7/8″ tubing is run in 5-1/2″ casing, conventional gas lift mandrels with 1.0" OD gas lift valves can be used. The preference would be to use a 7″ casing that would allow the use of 1.5″ OD gas lift valves. The 1.5″ OD gas lift valves have larger port options that would allow the wells to reach deeper lift depths through quicker unloading. This would minimize the time of multi-point lifting. The larger size of the casing could be considered based on the expected production profile of the well.

7.1.2.6 Surface Facilities

As unconventional wells flow with very minimal energy after the first 1–2 years, even a small amount of back pressure would hamper the production from these wells. The surface piping should be designed in such a way that back pressures are minimized, especially for gas lift application.

7.1.2.7 Ease of Switching the AL System

In order to maximize ultimate recoveries from these horizontal wells, the artificial lift methods selected must be flexible to cover the significant and rapid changes in flow rate over a well's life cycle. For this reason, different types of artificial lift often are implemented in the same well at different points in the production process. In fact, it is common for unconventional horizontal wells to use two or more forms of lift within only a few months of starting production.

The lift systems suitable for different phases of an unconventional well are shown in **Fig. 7-5**.

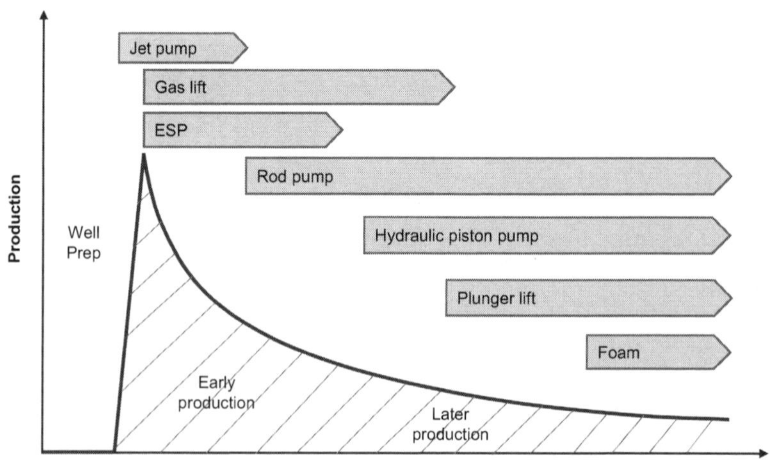

Fig. 7–5. AL Planning for unconventional wells
Courtesy: William Lane, *Completion and Artificial Lift Strategies for the Life of the Well*, Weatherford presentation in ALRDC [29]

7.1.2.8 Well Optimization Planning

Production optimization of an unconventional well starts with integrated planning. The main aspects that should be focused on are:

- **Well construction geometry**

  The well should be constructed such that lateral drainage should be easily possible. The pump should be landed in vertical or near-vertical sections (<70°) or straight tangents.

- **Completions**

  The completion should take into account the plan for future lift systems for the well. It should be possible to achieve change between lift systems without pulling out the completion. Subsurface sensors, if incorporated, help to optimize production from the well in an efficient manner.

- **Production optimization systems**

  Simple data-gathering systems should be incorporated to monitor the reservoir performance. Controls should be in place to make required changes. A data management system should handle the data from commissioning of the well.

In gas lifted wells, once the flow rates are down to the level that continuous flow is not working, intermittent flow will be set up. This can be accomplished with wireline intervention and surface modifications. Later, plunger lift or gas lift assisted plunger lift may be employed.

If the wells are completed with gas lift mandrels, the completion need not be pulled when changing into intermittent or plunger lift later.

7.1.2.9 Flexible System Optimization

In terms of system optimization, the facilities that are needed for the selected artificial lift method should be planned in advance and made available on time. This would be high pressure gas for gas lift, a clean power supply for ESP, and a hydraulic system for downhole jet pump. As the well production declines steeply, artificial lift should be adjusted accordingly. This will require adjustment of the gas lift pressure and volumes for gas lift wells, ESP speed using variable frequency drives (VFD), or pressure and volumes of power fluid for jet pumps, as needed to get the best out of the well at any time.

The other issues to deal with are the slugging problems in the surface gathering pipelines. Use of gas lift helps to maintain a better velocity; therefore, the impact due to low velocities and liquid holdup will be minimal with gas lift, especially with undulated surface lines. Other flow assurance issues such as wax deposition based on composition of oil and flowline temperature should also be addressed.

7.1.2.10 Achieving Flow Assurance in Unconventional Wells

Technology has been advancing to help operators continue production with high rate artificial lift in unconventional wells until it becomes necessary to switch over to another artificial lift.

In gas lift systems, efforts are made to attempt to inject lift gas below production packer so as to enhance fluid production from the horizontal sections.

In wells with long perforated intervals, insufficient velocities below the packer can cause liquid loading. Many traditional artificial lift methods are limited in their ability to address liquid loading issues in these types of wells. One solution that has been considered is to inject gas below the packer and deep into the perforated zones.

Advances in gas lift system design now provide several options for effectively recovering liquids from completion intervals below the packer. Options include Annular Bypass Assembly (ABA), Dip Tube, Enhanced Annular Velocity (EAV), Marathon AVE, and Dead String. Each employs unique techniques that allow for gas injection deep into the perforated zone, bypassing the lift gas through or around the packer.

The injected gas aerates the liquid in the perforated zone, decreasing the hydrostatic density and flowing gradient. This improves inflow from the formation. By also increasing velocity in the production string, these gas lift systems allow operators to more efficiently recover fluids and minimize or prevent liquid loading.[30]

Below-packer gas lift systems are being used successfully to economically produce wells in unconventional pays. The design and mechanics of each of the abovementioned systems should be carefully considered before deciding to implement a particular method.

ESP manufacturers have come up with an "unconventional extended-life ESP that comfortably operates at low flow rates and manages demanding production declines that are typical of an unconventional well." The details of such improvements should be evaluated thoroughly before installing in the well.

7.1.3 Conventional Gas Lift Design for Unconventional Wells

Gas lift for unconventional wells is designed differently from conventional wells. Standard nodal analysis models cannot be used for inflow description, as both the reservoir pressure and PI are changing. As a result, an "expected" pressure traverse cannot be fixed for creating the mandrel spacing design. This leads to major changes in the gas lift design approach for unconventional wells:

- A flowing gradient corresponding to a certain production rate may be used for the initial gas lift design, to start off with the mandrel spacing. This may be applicable for the upper section of the well. Below this level, it is customary to place mandrels at a bracket spacing of 500–550 ft.
- Dome-charged valves are used in all mandrel stations. Some operators prefer to use an orifice valve at the bottom-most mandrel. As the production declines, the flowing gradient will become lighter and the well will be able to lift deeper. There will be multi-point lifting till the well reaches the deepest gas lift mandrel.
- Injection pressure-operated unloading valves are normally used in these designs. The closely spaced mandrel stations installed in the bottom section help unload the well to the deepest mandrel. This is assisted by the decline of inflow that facilitates faster unloading.

- In a well that is 10,000 ft. deep (TVD), the design may have 12–18 mandrel stations, depending on the gas lift pressure available and the expected fluid gradient.

A sample mandrel spacing design for an unconventional producer is shown in **Fig. 7-6**. Each deeper IPO unloading valve is set to open at a slightly lower surface casing pressure. As the lower valve takes over injection of lift gas, the casing pressure gradually decreases and the upper valve closes. The unloading continues to a deeper mandrel.

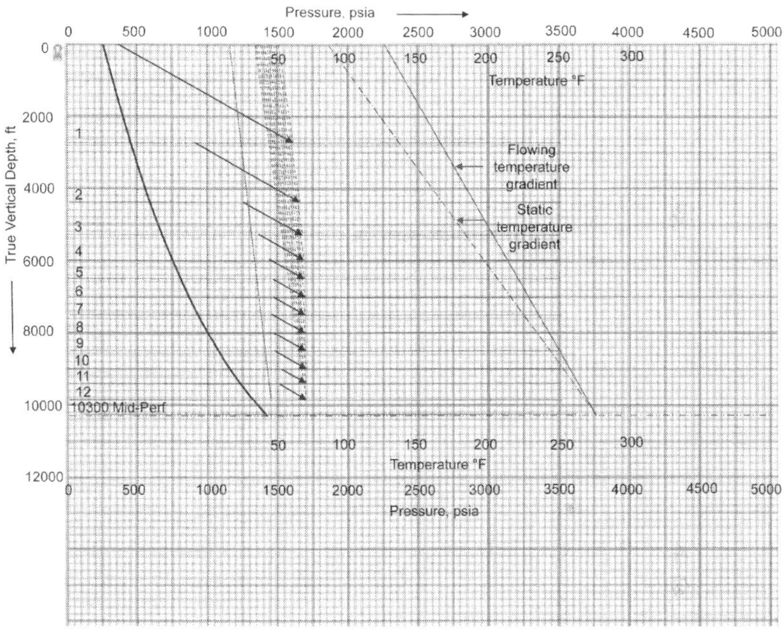

Fig. 7–6. Conventional mandrel spacing design for unconventional well

In this gas lift design, the mandrel spacing has been done for the upper section of the well using a heavier flowing gradient. This unloads the well to the fourth mandrel at 6000 ft. TVD. Beyond this depth, a bracket spacing of 500 ft. has been utilized till the 12th mandrel at 9800 ft. A gas lift kickoff pressure of 1500 psia is available.

When the flowing pressure gradient in the tubing becomes lighter due to decline of inflow (**Fig. 7-6**), the well would have unloaded to the bottom mandrel.

The golden rule that "multi-pointing is an inefficient use of lift gas and should be eliminated when possible" is true, but the achievement of this goal is based on the assumption that the gas lift valve has the capacity to flow the required amount of gas to achieve the desired GLR, given the upstream / downstream pressure conditions. While using gas lift early in the life of unconventional wells, the flow performance of the valve is much less than the required gas flow rate. This is due to the smaller port size of the 1" IPO valve and limited pressure differential available at depth. Consequently, a single valve is not capable of flowing the volume of gas needed to achieve single-point injection. The well is allowed to inject multi-point for a few weeks or months till the well unloads to the bottom mandrel.

Thus, trying to achieve single-point injection in an unconventional well while the reservoir pressure is declining rapidly should not be a design goal. It is not advisable to spend time or money to eliminate multi-pointing at the initial stages of production. When the reservoir pressure has declined to a fairly stable value, engineering efforts may be initiated to achieve single-point injection. (31)

7.1.4 Improved Gas Lift Design for Unconventional Wells

A new method has been used to achieve a better mandrel spacing for high-rate unconventional gas lift wells. This method uses a GLR approach, as shown in **Fig.** 7-7.

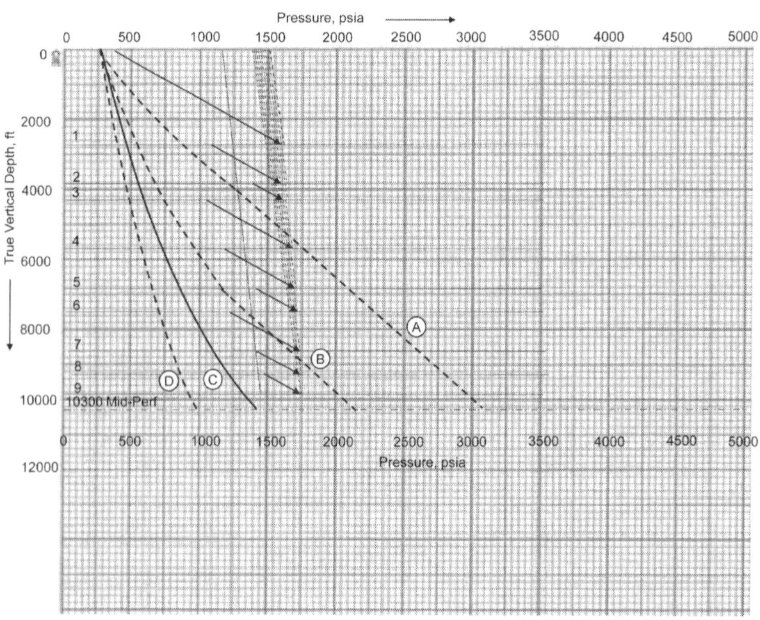

Fig. 7–7. GLR method of spacing design for unconventional well

- In the first instance, assuming gas lift would be commissioned for a 2000 BFPD rate, a TGLR curve (say, 500 scf/BBL) is generated for this rate. This is shown as *Curve A* in **Fig.** 7-7. The well can be lifted at about 4000 ft. under this condition. A close mandrel spacing to achieve this lift depth is designed. This requires three mandrels using the available lift gas pressure in this example. Injection-pressure-operated unloading valves will be installed in these mandrels.

- Next, a higher TGLR (1000 scf/BBL) curve is generated for a lower production rate, 800 BPD. This is shown as *Curve B*. The mandrel spacing will be continued to reach a suitable operating depth. This is achieved at about 7000 ft. in this case.
- Below this depth, the valve spacing can be far apart, as the tubing pressure at depth is not large enough, and the well will readily unload due to a lighter tubing gradient. *Curve C* shows the well unloaded to the bottom mandrel at 9800 ft. TVD.
- As the well inflow declines further, the gradient becomes lighter. This is shown as *Curve D*. After reaching the minimum gradient, gas lift may not be able to increase the drawdown further. The next step may be to switch to another low-rate lift method in the well.

The high-rate GLR design approach requires fewer gas lift mandrels. The transition from upper valves to lower valves can be greatly expedited on these unconventional wells by closely spacing the higher valves. The spacing using the GLR method allows the rapid unloading process to create a higher rate in the critical early time of production. It is at this point where the highest natural reservoir pressure is helping the lifting process, and the productivity index will be the greatest in the well's life.

Achieving a total fluid rate of 2500-plus BPD is well within the grasp of the high-rate unconventional GLR-based method. **Fig. 7-8** shows the data collected during the first 20 days of production with gas lift in a well from Southern Delaware Basin Texas. Based on data from an installed BHP gauge, accurate pressures have been tracked and production rates have reached more than 3000 BPD. A three-stage, 200-hp, gas-powered natural gas compressor with an operating discharge pressure of 1250 psig provided the required gas lift to this well. [30]

For optimization of gas lift in unconventional wells, the dynamic behavior of inflow should be considered as pseudo-steady-state, as steady-state inflow performance relationship (IPR) curves are not applicable. The shut-in bottom-hole pressure (SIBHP) may be calculated using rate transient analysis. Alternatively, dynamic nodal analysis may be used to estimate the SIBHP and PI to be used for optimization. This is required because both the reservoir pressure and PI decline in these wells.

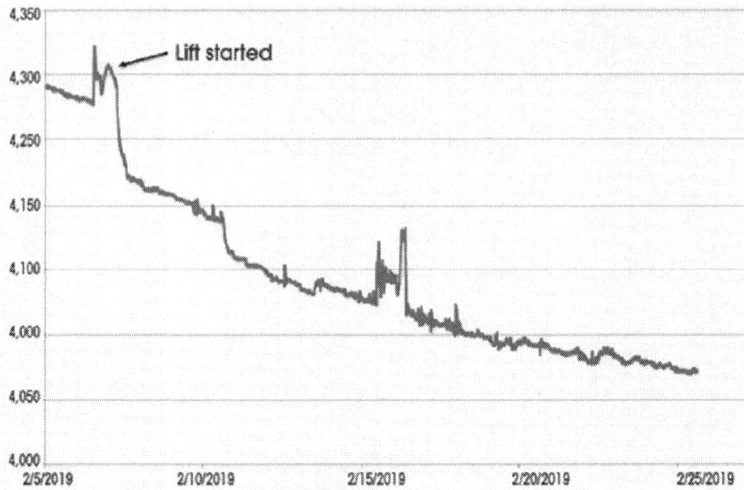

Fig. 7–8. Initial high-rate production with gas lift (BHP Data)
(Courtesy: Production Lift Companies and Concho Resources Inc.)

7.2 Optimization of Subsea GL Wells

Gas lift has been used in deepwater operations for more than two decades. The subsea production environment requires a different type of gas lift operation. Generally, the wellhead is located at the seabed. Production from several subsea Christmas trees are collected in a riser, and the fluids flow up the riser to an FPSO (floating production, storage, and offloading unit) or FPS (floating production system) at sea level.

- **FPSO**

  A floating production, storage, and offloading unit is a floating vessel used by the offshore oil and gas industry for the production and processing of hydrocarbons, and for the storage of oil.

- **FPS**

  Semi-submersible or ship-based system used to exploit the oil fields. The system has three main components: the floating production vessel, mooring system, and subsea system, which include the risers and well system. FPS processes and offloads hydrocarbons without storing them.

Gas lift is used in the deepwater environment mainly in two ways: the first is gas lifting the well to increase production, and the second is the riser base gas lift to prevent liquid loading in the riser from seabed to surface.

7.2.1 Single-point gas lift system

Unloading valves with nitrogen-charged bellows are used in traditional gas lift wells. Failure of the bellows would lead to multi-point injection, reducing the efficiency of the system. As the cost of intervention in a subsea well is considerably higher than for normal completions, a single-point gas lift (SPGL) system is used for deepwater oilfields. In the SPGL system, a single orifice is installed at a desired depth for well kickoff and continuous gas lift operation.

SPGL requires high gas lift pressures (2000–4000 psig) and large volumetric flow rates (5–12 MMscf/Day).

7.2.2 Deepwater Gas Lift Well Equipment

The objective of gas lift differs among deepwater wells:

- In some cases, production may commence initially on natural flow, and gas lift may be started later on.
- Sometimes the intent of gas lift is to provide a means of kickoff, and gas lift may be turned off once production is started.
- In some installations, continuous gas lift injection may be needed from the start of production.

Specific gas lift systems are needed for deepwater wells, considering the water depths, pressures, production rates, tubing sizes, gas lift injection pressures (kickoff and operating), temperature profiles, and metallurgy requirements.

Gas lift downhole equipment meeting subsea standards is now available from most of the suppliers. Gas lift valves rated for higher pressures with larger ports, equipped with barrier-certified check valves and reliable performance are needed for subsea operations. The higher pressure rating and pressure barrier features allow deeper depths of gas injection and ensure enhanced wellbore integrity. A few examples of new mandrels and valves are listed below:

- Schlumberger supplies XLift™ (high pressure) valves with the following improved designs:
  - 1.75″ OD valves, tested to 7500 psig, capable of injection rates up to 10 MMscf/Day
  - 1.5″ OD IPO valves, now redesigned to work with injection pressure up to 5000 psig (previously could work up to 2500 psig gas lift systems)
- The XLift valve also has a positive seal check valve, tested to subsurface safety valve (SSSV) standards, providing an additional barrier downhole. By using these valves, higher production rates can be achieved due to deeper injection depths. Furthermore, these valves provide improved tubing pressure integrity during all phases of operation.
- Weatherford supplies the DVX gas lift mandrels. Where H_2S production is likely, it is necessary to use a gas lift mandrel that would not allow any

produced fluids to enter into the casing annulus under any circumstances, even if the pocket is empty. The DVX mandrel has been developed to meet this criterion. External check valves are installed on the gas lift mandrel, and injection gas passes through these into the valve pocket area and the tubing string. The external checks prevent any flow from the tubing back into the casing, regardless of the type of valve or check mechanism, if any, loaded into the pocket. Valve changes can be done without any worry about the tubing-to-casing communication path.

- Petroleum Technology Company (PTC) supplies downhole gas lift valves, which could be relied upon to provide a "well integrity barrier" throughout the life of the well, thus maintaining the integrity of the primary well containment envelope.
- PTC also supplies actuated wellhead safety valves, located within the body of the wellhead, which will close in the event of damage to the HP lift gas flowline. This reduces the inventory of gas that could vent in the case of an HP gas lift line failure.

Many other service providers are available to design and supply customized equipment needed for complex deepwater and subsea completions.

7.2.3 Riser Base Gas Lift

In order to bring up production from wet Christmas trees located at the seabed to an FPSO or FPS at sea level, additional energy may be needed. In some locations, multiphase booster pumps are used to pump the produced fluids from subsea wells to the surface. Riser base gas lift may be used instead, if the fluid velocities, fluid properties (GOR, viscosity, water cut, etc.), and temperature are in favor.

Fig. 7-9 shows the arrangement of a subsea asset that has oil production, water, and gas injection and dedicated gas lift risers for riser base gas lift.

When dealing with multi-phase flow in subsea flowlines and risers, a wide range of flow assurance issues come into play. These include hydrate formation, wax deposition, asphaltenes, sand and solids transport, corrosion, erosion, scale deposition, and slugging in risers.

Riser base gas lift involves injection of a predetermined rate of gas into the production flowline (or riser) at the sea floor. The main objectives of riser base gas lift are:

- **Production enhancement by reduction of hydrostatic gradient**

  Gas lift is normally used for production enhancement if water cut is high and/or the reservoir fluid GOR or reservoir pressure is low. Gas injected into the riser base continuously lifts liquid out of the riser, preventing buildup of liquid. Lift gas decreases the liquid holdup in the riser, and thus reduces the hydrostatic pressure in the flowline. [33]

Fig. 7–9. Subsea field arrangement at seabed
Courtesy: M. A. Usman1, A. O Olatunde, T. A. Adeosun and O. L. Egwuenu [32]

- **Flow stabilization by slug suppression and velocity improvement**
  Severe slugging may cause shutdown of the surface facilities. Riser base gas lift increases the superficial velocity of gas, which diminishes the severity and cycle time of the slugging condition. If enough gas can be injected at the riser base to change the flow inside the riser to hydrodynamic slug flow or churn flow, the severe slugging problem can be mitigated. The flow will become very stable if annular flow is achieved, but this may require a significant amount of lift gas injection and it may not be practical.
- **Flowline depressurization to prevent hydrate formation**
  This is required to prevent hydrate formation after an unexpected shutdown. If not depressurized, hydrates will form and plug up the lines, as hydrate formation is favored by conditions of high pressure and low temperature.

Additional problems due to Joule-Thomson cooling of the injected gas may be encountered in riser base gas lift. Lift gas may cause cooling and make the flow conditions more susceptible to wax precipitation and hydrate formation. Different methods of gas injection are now available to minimize J-T cooling. The system should be designed based on typical flowing conditions expected in the field.

Fig. 7-10 shows how a riser base gas lift helps to maintain stable flow and increase production from a subsea field. Once gas is cut off, the average flowline pressure increases by about 20 bar, severe slugging sets in, and the flowline temperature decreases by about 25°C. (33)

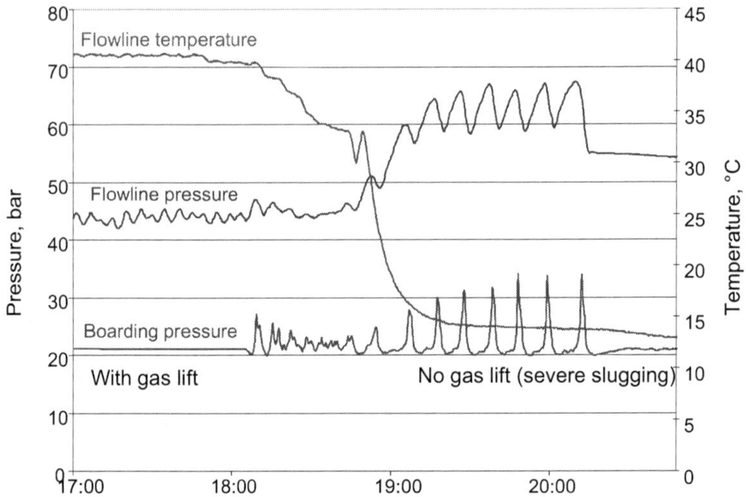

Fig. 7–10. Gas lift for slug suppression and flow stabilization
(Courtesy: OTC-18820-MS, Subash S. Jayawardena, George J. Zabaras and Leonid A. Dykhno) (33)

8

Data Analytics and System Optimization

The science of data analytics has already impacted the society in many areas. The energy industry is in the process of making use of advanced data analytics tools such as artificial intelligence, machine learning, and neural networks. The area of production system optimization has a tremendous opportunity to use big data analytics, as plenty of data is available from the various locations of operating assets. The goal is not only to predict failures, but to go one step further: to suggest actions to prevent failures in advance. This is the future of production optimization.

This chapter discusses the following topics:

- Introduction to concepts of machine learning, artificial intelligence, and neural networks
- Application of emerging data science methods in data processing, optimization, and automation
- Use of statistical models to detect deviation from normal before downtime occurs
- Application of big data analytics to maximize asset performance and improve profitability
- Use of hybrid models for system optimization

8.1 Data Analytics and Machine Learning

Data analytics, machine learning, and artificial intelligence have been applied successfully in many parts of businesses such as finance and healthcare. The energy industry is poised to witness a much bigger revolution in the near future by using data-driven analysis to improve reliability and predictability. Numerical and statistical analysis of real-time data could provide clear indications of sub-optimal performance of wells, equipment, and pipelines, and alert operations before any downtime is caused.

A few terminologies associated with data analytics and machine learning are described first. This is followed by some insights into how these emerging technologies can be applied for surveillance, automation, and failure prediction in the oilfield.

8.1.1 Definition of Big Data

The term *big data* refers to large, unstructured datasets with non-standard formats or frequency of data capture. Two aspects that should be highlighted are:

- The volume of data in the first place, from a few terabytes (TB) to many petabytes (PB). The data becomes large because of two reasons: first, data is being captured from several process points (wells, pipelines, rotating machineries, process plants, and storage areas); second, data is now available at increased capturing frequency.
- Other than the volume, it is structured and unstructured data available in many formats.

Such datasets often combine what is called "data in motion" (e.g., real-time data) with the configuration and historical data called "data at rest." Analytics can be used to analyze the past, evaluate current performance, and predict future behavior of a system or process.

Big data has three main features: volume, velocity, and variety. [34]

- Volume—the size of the data
- Velocity—the rate at which the data is changing, or how often it is created
- Variety—different types and formats of data, also different ways of analyzing the data

In a practical sense, big data is difficult to deal with, mainly because of the variety of unstructured data. It requires proper storage, management, integration, federation, cleansing, processing, and analyzing.

The industry needs to establish a common framework or road map for big data management, especially on issues such as data format and cybersecurity issues around gathering, storing, and sharing information. This would encourage more operators to adapt big data analytics and leverage on digital transformation of their businesses.

8.1.2 Cloud Storage

Cloud storage allows users to save data and files in an offsite location that can be accessed either through the public internet or a dedicated private network connection. The data is transmitted and stored on remote storage systems, where it is maintained, managed, backed up by a third-party cloud provider, and made available to users over a network—typically, the internet.

Storage services are provided on demand in the cloud, with capacity increasing and decreasing as needed. Organizations opting for cloud storage eliminate the need to buy, manage, and maintain in-house storage infrastructure. Users generally pay for their cloud data storage on a per-consumption, monthly rate. In general, cloud storage has radically driven down the per-gigabyte cost of storage.

The most commonly used cloud services are from AWS (Amazon Web Services), Microsoft Azure, and Google Cloud.

8.1.3 Definition of Data Analytics

Data analytics has been defined as "a set of techniques and tools intended for data access, integration, analysis, and visualization, which makes it possible to identify valuable patterns to improve the decision-making process in a work environment." [33] In other words, data analytics is the process of applying algorithms in order to analyze sets of data and extract useful and unknown patterns, relationships, and information so that anything that shows a deviating trend from normal can be recognized when the deviation starts happening.

8.1.4 Types of Data Analytics

There are four types of data analytics:

- **Descriptive analytics** is a process of describing historical trends in data. It summarizes the data in a meaningful and descriptive way. The process involves collection of relevant data, processing of the data, data analysis, and data visualization. This process provides essential insights into past performance and explains what happened. [35]
- **Diagnostic analytics** helps answer questions about why things happened. Analysts take the findings from the descriptive analytics and dig deeper to find the cause. The performance indicators are further investigated to discover why they got better or worse.
- **Predictive analytics** uses past experience to predict the future. Historical data is used to identify trends and determine if they are likely to recur. A variety of statistical and machine learning techniques such as neural networks, decision trees, and regression are used to predict the future behavior of a process or system, based on the analysis of historical data.
- **Prescriptive analytics** allows businesses to make informed decisions in the face of uncertainty. By using insights from predictive analytics, data-driven decisions can be made. Prescriptive analytics techniques rely on machine learning strategies that can find patterns in large data sets. By analyzing past decisions and events, the likelihood of different outcomes can be estimated. Prescriptive analysis offers suggestions on what should be done.

8.1.5 Steps of Data Analytics

The four basic steps in the data analytics process are:

- **Data mining**: This is usually the most time-intensive step in the data analysis pipeline. It involves extracting data from unstructured data sources. The data may be in any form—written text, large complex databases, or raw sensor data. The step requires the analyst to extract, transform, and load (ETL) data, thus converting raw data into a useful and manageable format.

- **Data management or data warehousing**: This involves designing and implementing databases that allow easy access to the results of data mining. This step generally involves creating and managing SQL databases. Non-relational and NoSQL databases are also becoming common now. There are many databases available in the market such as MongoDB, Oracle, and PostgreSQL that provide high performance, fast processing, and scalability.

  Note:
  - A database is a systematic collection of data. SQL stands for Structured Query Language, which is a computer language for storing, manipulating, and retrieving data stored in a relational database. SQL is the standard language for a relational database system.
  - NoSQL, which stands for "not only SQL," is an alternative to traditional relational databases. NoSQL is an approach to database design that can accommodate a wide variety of data models, including key-value, document, columnar, and graph formats. NoSQL databases are particularly useful for working with large sets of distributed data.
- **Statistical analysis**: This is the heart of data analytics. Both statistical and machine learning techniques are used to analyze the data. Big data is used to create statistical models that reveal trends in the data. These models can then be applied to new data to make predictions and inform decision-making. Statistical programming languages such as R or Python are essential to this process.
- **Data presentation**: This final step in data analytics allows insights to be shared with stakeholders. Data visualization is a critical aspect of the presentation. Compelling visualizations are essential to convey the story in the data for executives and managers to understand the importance of these insights. [36]

Many data visualization and analytics tools are available in the market. Example tools are Tibco Spotfire, Tableau, and Power BI.

8.1.6 Machine Learning

Machine Learning (ML) is an application of artificial intelligence (AI) that provides systems the ability to automatically learn and improve from experience without being explicitly programmed. ML focuses on the development of computer programs that can access data and use it to learn for themselves. [37]

How ML differs from traditional programming can be easily understood from **Fig. 8-1**. In traditional programming, data and program are run on the computer to produce the output. But in ML, data and output are run on the computer to create a program. This way, the program is able to learn and improve its output by the learning process. (38)

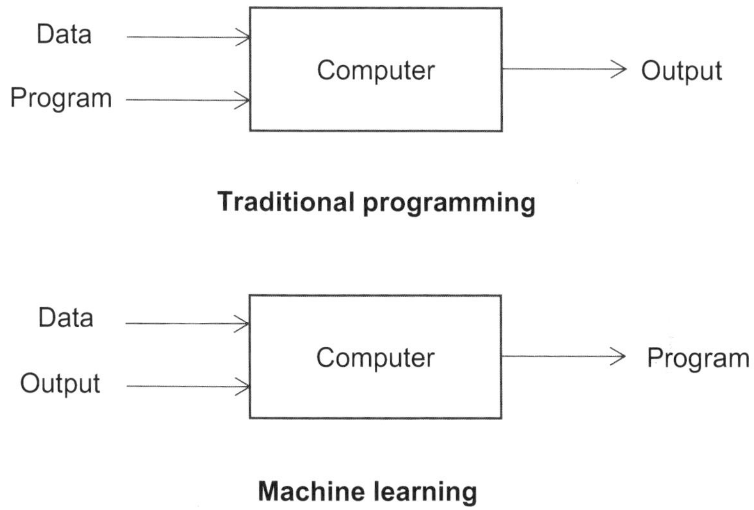

Fig. 8–1. Traditional programming vs. machine learning

The process of learning begins with observations of data, such as examples, direct experience, or instruction, in order to look for patterns in the data and to make better decisions in the future based on the examples that are provided. The primary aim is to allow the computers to learn automatically without human intervention or assistance and to adjust actions accordingly. Deep learning techniques enable this automatic learning through the absorption of huge amounts of unstructured data such as text, images, or video.

ML algorithms are often categorized as supervised, unsupervised, or semi-supervised:

- **Supervised machine learning algorithms** can apply what has been learned in the past to new data, using labeled examples to predict future events. Starting from the analysis of a known training dataset, the learning algorithm produces an inferred function to make predictions about the output values. The system is able to provide targets for any new input after sufficient training. The learning algorithm can also compare its output with the correct, intended output and find errors in order to improve the model accordingly.
- **Unsupervised machine learning algorithms** are used when the information used to train is neither classified nor labeled. Unsupervised

learning studies how systems can infer a function to describe a hidden structure from unlabeled data. The system does not figure out the right output, but it explores the data and can draw inferences from datasets to describe hidden structures from unlabeled data.

- **Semi-supervised machine learning algorithms** fall somewhere in between supervised and unsupervised learning, since they use both labeled and unlabeled data for training—typically a small amount of labeled data and a large amount of unlabeled data. The systems that use this method are able to considerably improve learning accuracy.

ML enables analysis of massive quantities of data. While it generally delivers faster, more accurate results in order to identify profitable opportunities or dangerous risks, it may also require additional time and resources to train it properly. Combining ML with AI and cognitive technologies can make it even more effective in processing large volumes of information, to arrive at suggested action plans.

Fig. 8-2 summarizes the various steps in building an ML model and training it for performing the required function.

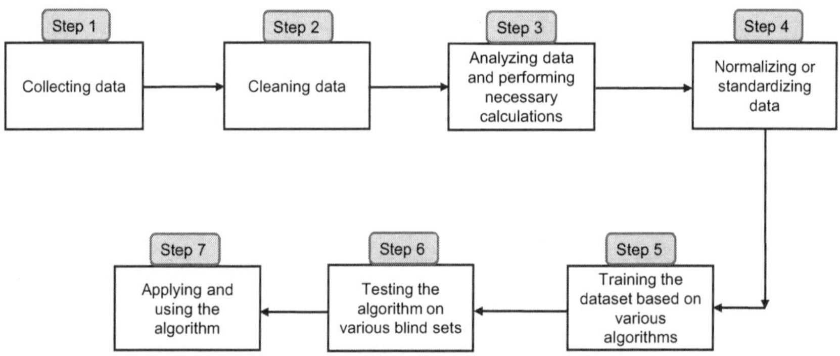

Fig. 8–2. Building a ML model

1) Data visualization:

   Distribution plots, box plots, and scatter plots are used to visualize the data. The ML engineer will ensure that the distribution plot of each parameter is normal and, otherwise, will adjust it using various techniques.

2) Anomaly Detection:

   After data visualization, anomalous data points are investigated to make sure all data points are valid. When validity is doubtful, the points may be removed.

3) Collinearity Removal:

   Pearson correlation coefficient is used to remove co-linear parameters that have more than 90% collinearity. This will eliminate parameters that essentially provide the same information. Pearson correlation coefficient is a measure of the strength of a linear association between two variables. The correlation attempts to draw a line of best fit through the data of two variables. The coefficient can take a range of values from +1 to -1.

- A value of 0 indicates that there is no association between the two variables.
- A value greater than 0 indicates a positive association; that is, as the value of one variable increases, so does the value of the other variable. The Pearson correlation coefficient indicates how well the data points fit this new model/line of best fit. For example, in a positive association, a coefficient of 0.8 indicates less variation around the line of best fit.
- A value less than 0 indicates a negative association; that is, as the value of one variable increases, the value of the other variable decreases. A coefficient of -0.3 indicates that there is too much variation around the line of best fit.

4) Feature Ranking and Selection:

   Algorithms such as random forest and extra trees may be used to rank the features and potentially remove the ones that have the lowest impact on the model.

5) Normalization / Standardization:

   The data may be normalized to be between 0 and 1 or -1 and 1 by using minimum or maximum scaler to normalize the data. Standard scaler may be used to standardize the data to have a mean of 0 and standard deviation of 1.

6) Finalization of the Model Algorithm:

   The algorithm for the ML model will be decided by the data scientist. The selection of model algorithm is based on accuracy, precision, and recall, so that a better result is obtained with training and test data.

7) Modification with Learning:

   As the algorithm will be tested on data sets and based on test results, the model will modify the algorithm. As the application continues, the ML model will learn and improve itself to give more accurate results, as required.

8.1.7 Artificial Intelligence

Artificial intelligence (AI) refers to the simulation of human intelligence in machines that are programmed to think like humans and mimic their actions. The goals of artificial intelligence include imitating human cognitive activity. The term is also applied to any machine that exhibits traits associated with a human mind such as learning and problem-solving. [39]

The ideal characteristic of artificial intelligence is its ability to rationalize and take actions that have the best chance of achieving a specific goal.

Some people believe that innovators may soon be able to develop systems that will exceed the capacity of humans to learn or reason out any subject. However, others remain skeptical about this because all cognitive activity is laced with value judgments that are subject to human experience.

8.1.8 Neural Network

A *neural network* (NN) is a series of algorithms that endeavor to recognize underlying relationships or patterns in a set of data, through a process that mimics the way the human brain operates.

Essentially, neural networks are composed of layers of computational units called neurons, with connections in different layers. These networks transform data until they can classify it as an output. Each neuron multiplies an initial value by some weight, sums results with other values coming into the same neuron, adjusts the resulting number by the neuron's bias, and then normalizes the output with an activation function. [40] This is shown in **Fig. 8-3.**

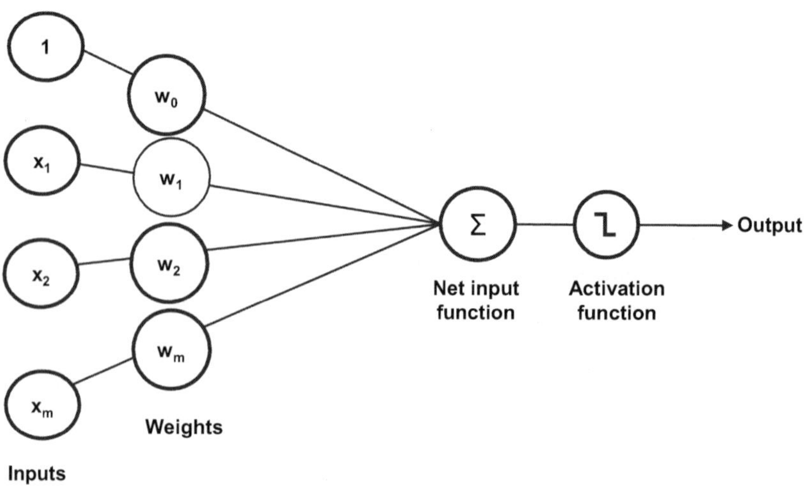

Fig. 8–3. Output from neural network

Neural networks generally perform supervised learning tasks, building knowledge from data sets where the right answer is provided in advance. The networks then learn by tuning themselves to find the right answer on their own, increasing the accuracy of their predictions.

To do this, the network compares initial outputs with a provided correct answer, or target. A technique called a cost function is used to modify initial outputs based on the degree to which they differed from the target values. Finally, cost function results are then pushed back across all neurons and connections to adjust the biases and weights using a push-back method called *back-propagation*. This iterative learning process is the most interesting thing about neural networks.

The network processes the records in the "training set" one at a time, using the weights and functions in the hidden layers, then compares the resulting outputs against the desired outputs. Errors are then propagated back through the system, causing the system to adjust the weights for application to the next record. This process occurs repeatedly as the weights are tweaked. During the training of a network, the same set of data is processed many times as the connection weights are continually refined.

Neural networks rely on training data to learn and improve their accuracy over time. However, once these learning algorithms are fine-tuned for accuracy, they are powerful tools in computer science and artificial intelligence. One of the most well-known neural networks is Google's search algorithm. It is true that neural networks have made modern computer systems more useful by making them more human.

The advantages of neural networks include their high tolerance to noisy data, as well as their ability to classify patterns on which they have not been trained. This aspect of neural networks can be efficiently used for monitoring gas lift wells and networks.

The interconnection among the various data science techniques is shown in **Fig. 8-4**.

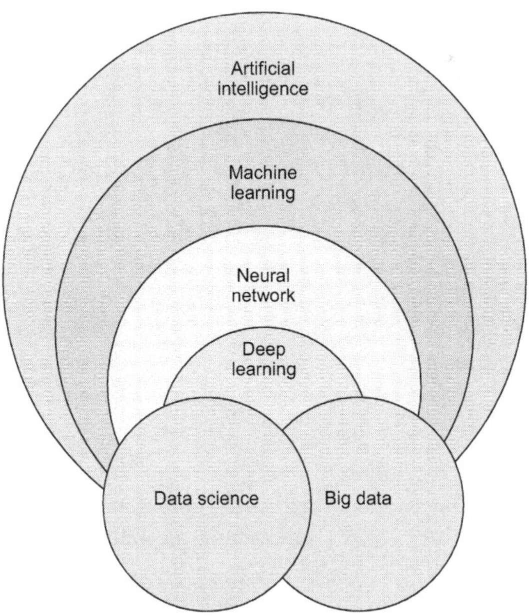

Fig. 8–4. Interconnections between various areas of data science

8.1.9 Industrial Internet of Things

The *Internet of things* (IoT) describes a network of physical objects that are embedded with sensors and software for the purpose of connecting and exchanging data with other devices and systems over the internet. A "thing" in the IoT can be, for example, a person with a heart monitor implant, or an automobile that has built-in sensors to alert the driver when tire pressure is low. The IoT aims to enable the interconnection and integration of the physical world and cyberspace.

The *Industrial Internet of Things* (IIoT) is defined as machines, computers, and people enabling intelligent industrial operations, using advanced data analytics for transformational business outcomes. The connectivity in the IIoT allows for data collection, exchange, and analysis, potentially facilitating improvements in productivity and efficiency, as well as other economic benefits in the industrial environment.

The IoT and the IIoT both connect devices to the internet and make them smarter. The difference is that while the IoT works to make the lives of people more convenient and easier, the IIoT works to increase safety and efficiency in any industrial environment. Stated differently, the IoT focuses on B2C (business-to-consumer) and the IIoT is B2B (business-to-business). The idea behind machine-to-machine communication in the IIoT is to reduce human interventions as much as possible and achieve the highest level of automation. Handheld devices are used in a plant to communicate with machines to know their health status.

In summary, the key benefits of the IIoT in the context of an industry are improved and intelligent connectivity between devices or machines, increased efficiency, savings of cost and time, and enhanced industrial safety. (41)

8.2 Application of Data Analytics to the Oilfield

Some elementary methods of using data science for processing real-time data from a gas lifted field are covered in this section. The goal is to use the DA / ML technologies to address real-life challenges in the surveillance and optimization of gas lift wells and systems. The data collected should be accessed by a data scientist who can explore ways of analysis after he understands the gas lift operation of the field. Depending on the process and the actual data available, analysis possibilities will be explored.

Listed below are a few of the real-life tasks in gas lifted fields:

Surface issues

1) Validating or accepting a new production well test from a gas lifted well
2) Identifying gas lifted wells that are underperforming

3) Predicting freezing problems in well gas lift line or choke
4) Monitoring well slugging that affects production from other wells connected to the same header
5) Predicting wax or scale buildup tendency in surface flowlines
6) Spotting liquid buildup in horizontal / inclined gas lines, causing high pressure drop
7) Allocating optimum lift gas in a closed network of a gas lift system
8) Identifying flow assurance issues in long multi-phase production pipelines

Downhole issues

1) Effect of casing-tubing communication ("popped up" gas lift valves or a hole in tubing)
2) Well flow assurance: instability or slugging in horizontal / deviated / vertical sections
3) Well downhole plugging (deposition of salt, scale, asphaltene, or paraffin, etc.)
4) Tubing metal loss due to corrosion- / erosion-related issues

8.2.1 Exploratory Data Analysis

Before proceeding to analysis, the data file should be read easily for the analysis in Python using Panda library. This is called exploratory data analysis (EDA). The objectives are:

- Cleansing the data
- Understanding the data
- Features scaling or ranking which column is important
- Identifying data anomaly (if any)

A sample data set from an offshore gas lift field will be considered. A part of the data is presented in **Table 8-1**. The data has critical parameters taken from gas compression, production platform, and separation facilities.

The data is extracted and compiled into a **csv** file. A five-point statistical summary of gas lift variables is created with the simple "describe" function (**Table 8-2**). Further analysis of the data helps the analyst to understand the spread of numerical data. The minimum value, mean values, different percentile values, and maximum values are calculated, as listed in the table.

Table 8–1. Gas lift well and system data

Gas compressor station data					PF gas lift manifold		Individual well data (Example Well A1)					PF prod manifold		Prod separator facilities			
Compr suction press	Compr suction temp	Compr disch header press	Compr disch header temp	Compr disch flow rate	GL manifold press	GL manifold temp	Well GL inj pressure	Well GL inj temp	Well GL inj rate (Set Point 2.0)	Flowing wellhead press	Flowline temp	Prod manifold press	Prod manifold temp	Prod sep inlet press	Prod sep inlet temp	Prod sepr press	Prod sepr temp
psig	DegF	psig	DegF	MMscf/D	psig	DegF	psig	DegF	MMscf/D	psig	DegF	psig	DegF	psig	DegF	psig	DegF
115	110	1380	124	90.0	1278	82	1030	82	2.1	208	160	190	155	135	110	130	105
114	109	1382	124	88.0	1280	83	1028	82	2.1	206	160	192	158	135	110	130	105
114.5	110	1380	125	90.0	1278	83	1025	82	2.0	208	160	190	156	135	110	130	105
113.8	109.5	1383	125	89.0	1281	83	1029	82	2.1	208	160	188	155	134	110	130	105
114.8	110	1376	125	90.0	1274	83	1031	82	1.9	205	160	190	155	135	110	130	105
114	110	1379	125	90.5	1277	84	1032	82	2.1	208	160	188	158	135	110	130	105
114.5	110.5	1380	125	90.0	1278	83	1030	82	2.1	208	160	190	156	135	110	130	106
113.8	110	1380	125	88.0	1278	84	1028	82	1.9	209	159	192	155	135	110	130	106
114.5	110	1382	125	90.0	1280	84	1030	82	2.0	210	160	190	155	134	110	130	106
115	111	1380	125	90.0	1278	85	1024	83	2.0	208	160	187	158	135	110	130	106
114.5	110.5	1380	125	89.0	1278	85	1031	83	2.1	206	160	190	156	135	110	130	106
113.8	110	1378	126	90.0	1276	85	1032	83	2.1	208	160	190	155	133	110	130	106
114.5	112	1380	126	90.5	1278	85	1030	83	2.0	208	160	190	155	135	110	130	106
115	111.5	1380	126	90.0	1278	85	1028	83	2.1	205	160	190	158	135	110	130	106

Table 8–2. Numerical spread of data

```
df[df.Well == 'A3'].describe().round(decimals=2)
```

	PressurePFAGLManifold	GLInjPressure	GasLiftInjRate	WellHeadPressure	WellHeadTemp
count	360.00	360.00	360.00	360.00	360.00
mean	1277.83	995.23	1.88	208.79	155.02
std	1.44	96.00	0.37	5.19	14.13
min	1274.00	740.00	0.50	195.00	118.00
25%	1278.00	1029.00	1.90	208.00	160.00
50%	1278.00	1032.00	2.00	210.00	160.00
75%	1278.00	1036.00	2.10	213.00	161.00
max	1281.00	1040.00	2.15	215.00	162.00

8.2.2 Identification of Outliers

Box plots are a standardized way of displaying the distribution of data based on a five-number summary—minimum, first quartile (Q1), median, third quartile (Q3), and maximum. The outliers are usually identified with the help of a box plot.

Outlier refers to a person or thing differing from all other members of a particular group or set. The box plot shows some of the data points outside the range of the data (**Fig. 8-5**). Review of the plot shows that wells A2 and A4 have a few outlier data points. This could be indicative of the presence of a well slugging problem during that period.

```
var = 'Well'
data = pd.concat([df['GasLiftInjRate'], df[var]], axis=1)
f, ax = plt.subplots(figsize=(12, 5))
fig = sns.boxplot(x=var, y="GasLiftInjRate", data=data)
```

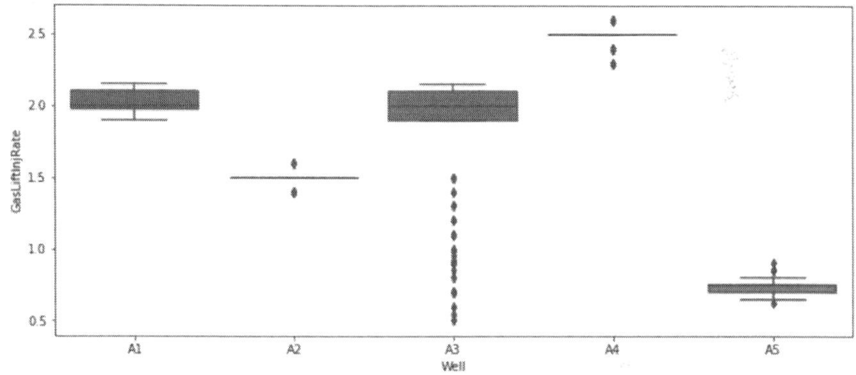

Fig. 8-5 Box plot

8.2.3 Searching for Dependent Variables

A *heat map* is a type of plot used to find dependent variables. This is a simple method to find the relationship between two data values. In **Fig. 8-6**, positive correlation is represented by lighter shades and negative correlation by darker shades. The target variable gas lift injection rate shows a strong relationship with other dependent variables, i.e., gas lift injection pressure, flowing wellhead pressure, and temperature.

```
#correlation matrix
k = 6 #number of variables for heatmap
corrmat = df.corr()
cols = corrmat.nlargest(k, 'GasLiftInjRate')['GasLiftInjRate'].index
cm = np.corrcoef(df[cols].values.T)
sns.set(font_scale=1.25)
hm = sns.heatmap(cm, cbar=True, annot=True, square=True, fmt='.2f', annot_kws={'size': 12},
                 yticklabels=cols.values, xticklabels=cols.values)
plt.show()
```

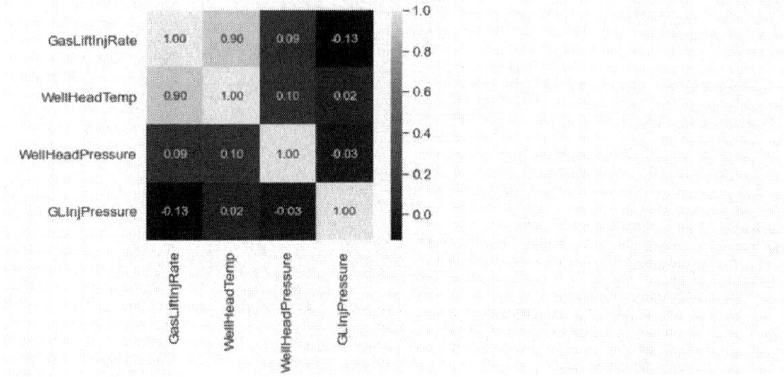

Fig. 8–6. Heat map

8.2.4 Locating Abnormal GL Well Performance

Having identified the close relationship between gas lift injection rate and gas lift injection pressure, the data set is plotted for all wells in **Fig. 8-7**. Data from Well A5 appears in the top left corner. The well has a low gas lift injection rate, with high gas lift injection pressure. The well appears to have a gas lift unloading problem.

```
var = 'GasLiftInjRate'
data = pd.concat([df['GLInjPressure'], df[var]], axis=1)
data.plot.scatter(x=var, y='GLInjPressure', ylim=(0,1400));
```

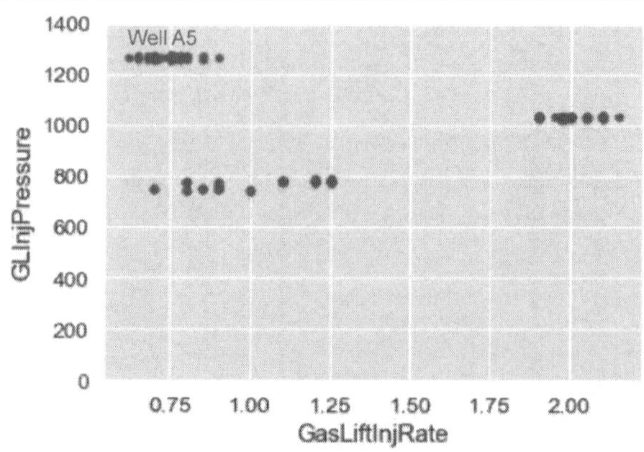

Fig. 8–7. GL injection rate vs. pressure

8.2.5 Anomaly Detection

This feature can be used to detect anomalies within the gas lift parameters of wells. Normally, alarms are set in real-time systems to alert operations when a parameter deviates from normal. These are set based on a "normal" operational value. In reality, the operating parameters might change over time, and therefore, a statistical approach is used to set *dynamic set points* as operations continue.

The steps to detect anomalies in data are:

- Remove the outlier from the data set
- Calculate the rolling average/median (define the window size for this)
- Calculate the rolling standard deviation
- Set the lower and upper limit standard deviation (define 3 sigma lower or upper)

The outlier data points present in the data can cause errors in calculation of rolling average and standard deviation. Hence the outlier is removed from the data set first. This is part of data handling profiling or exploratory data analysis. The data points outside the lower and upper whiskers in a box plot are considered as outliers. The outlier data points are removed using an IQR (inter-quartile range) technique.

```
# Removing ouliters from datasets using IQR method
Q1=df.WellHeadPressure.quantile(.25)
Q3=df.WellHeadPressure.quantile(.75)
IQR=Q3-Q1
lower_limit_WellHeadPressure=Q1-1.5*IQR
upper_limit_WellHeadPressure=Q3+1.5*IQR
lower_limit_WellHeadPressure,upper_limit_WellHeadPressure
df_no_outlier= df[(df.WellHeadPressure>lower_limit_WellHeadPressure)&(df.WellHeadPressure<upper_limit_WellHeadPressure)]
```

Once the data is cleaned, the rolling average or median and standard deviation are calculated for the window size. In order to define the operating envelope, from the mean, 3 sigma standard deviation is calculated. The window size is 24-hour data points, i.e., daily window. The operating envelope or the window size is adjusted or fine-tuned as per the well operating conditions.

```
#Defining dynamic operating envelope
df_no_outlier['lower_limit_WellHeadPressure'] =
df_no_outlier.WellHeadPressure.rolling(window=24).mean() -3*df_no_outlier.WellHeadPressure.rolling(window=24).std()
df_no_outlier['upper_limit_WellHeadPressure'] =
df_no_outlier.WellHeadPressure.rolling(window=24).mean() +3*df_no_outlier.WellHeadPressure.rolling(window=24).std()
```

The real-time data points falling outside the envelope are detected points of anomaly. This can be statistically set by the algorithm for the field operation team based on the historical data (**Fig. 8-8**).

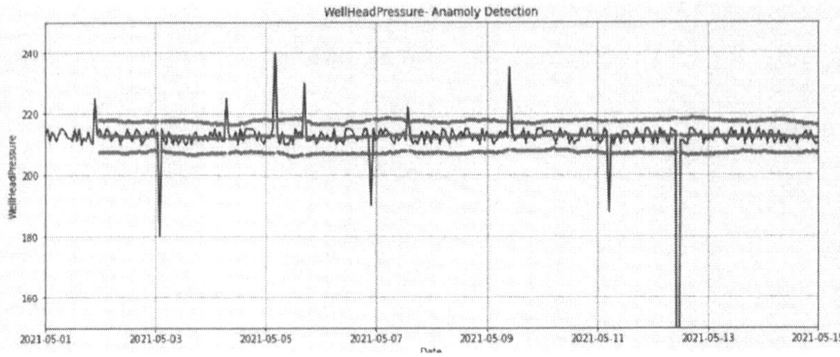

Fig. 8–8. Anomaly detection

8.3 Data Analytics Applied to GL Systems

8.3.1 Compressor Maintenance and Reliability

Gas compression is a major cost item of a gas lift operation. By improving the reliability of the compressors, significant production loss can be avoided.

Improving the reliability of gas compressors was one of the earliest tasks executed successfully using data analytics. In the beginning stage, the focus was on *preventive maintenance,* using condition-based monitoring. With the machines being located in a physically accessible location ("compressor house"), fixing various probes and collecting and transmitting continuous data was not difficult.

Sensors measure critical parameters like temperature and vibration and detect conditions that could lead to equipment failure. The systems initiate alarms when the operating parameters deviate from normal. This provides a chance for people to inspect and rectify minor abnormalities. If this is not rectified, then the machine will shut down. Alarm and shut systems are put in place to avoid permanent damage to the machineries.

With a wealth of experience, combined with technical support from compressor manufacturers, the operator implements the preventive maintenance program. Significant improvement in reliability and cost savings in equipment maintenance are achieved as a result of the program. Importantly, the outcome is tangible and easily justified.

This concept is then progressed further to *predictive maintenance* by applying data analytics to predict potential equipment failures, so that the failures do not come as a surprise. This helps the operator plan actions to manage the production loss. Knowing that the gas turbine is getting fouled, the operator prepares for the maintenance and could order the service and replacement parts in advance. This helps avoid the unplanned shutdown of critical equipment like the compressors. Unplanned downtimes are very expensive in gas lifted fields due to the associated production loss and the reactive nature of the response that will be needed to correct the situation.

Now, with the use of smart technologies, ML and AI models are able to prescribe actions that could be taken to sustain field operation efficiency even if degradation of performance has been observed. This allows the operator to move from predictive to *prescriptive maintenance*. The systems identify the critical equipment needed for the process and suggest the inventory that should be kept to manage the field efficiently, thus helping to keep the inventory costs within limits. This also assists in forward planning of scheduled maintenance activities and engine overhauls.

8.3.2 GL Wells and Network

At a broad level, data comes from three major sources: wells, pipelines, and the process plant. The process plant consists of vessels like separators, heat exchangers, and de-salting towers and running machineries like pumps and compressors.

With regard to well data, data may come from three different categories of wells:

1. Wells with simple surface instrumentation only
2. Wells with surface data and minimum downhole data, e.g., a permanent downhole gauge (PDG)
3. Wells with surface data and a full suite of continuous downhole surveillance and control (intelligent well)

DA and ML algorithms process all forms of data available from the surface and sub-surface of the entire asset. They can identify any trend that seems to deviate from a normal behavior, to be recognized as soon as the deviation starts happening. All the available data will be processed and analyzed to find hidden trends or patterns. Using the patterns observed during past incidents, future abnormalities can be predicted.

8.3.3 Well Test Validation

Performance monitoring of a naturally flowing or artificial lifted oil well can be executed with the help of production well tests. Even if other parameters suggest the well is normal, a production well test offers confirmation for the actual volume of fluids produced by the well. A well test validation process should be put in place to ensure the accuracy of these well tests.

Depending on the testing facilities available, operators test the wells as frequently as possible. In large fields with many test separators, a well may be tested once a week or fortnight. In some assets, it takes several months for a well to get tested.

With the use of advanced sensors, instrumentation, and data communication technology available at the wellhead and in the process plant, the well test data gathered is entered into the system once the test is completed in the field. Traditionally, production engineers review the well test data. However, if there are many tests done every day, the engineer may not have sufficient time to review the background of all the wells tested on a day. At the first level, it is thus useful to have some smart technology that can organize, filter, and analyze the data and confirm whether the test data is good or not. It will be valuable to quickly identify if the test is suspect or if the well production is too low.

A normal production well test from a gas lifted well may have the following parameters:

Measured values

1) Liquid rate
2) Water cut
3) Total gas rate
4) Flowing wellhead pressure (FWHP), oil manifold pressure
5) Casing head pressure (CHP, injection pressure), gas lift manifold pressure
6) Gas lift injection rate (Q_{gi})
7) Flowline temperature
8) Production choke size
9) Gas lift choke size

Calculated values

10) Oil rate, water rate
11) TGLR, injection gas-liquid ratio (IGLR), produced GOR, IGOR
12) Gas lift injection depth (gas lift design data, validated with models)
13) Reservoir pressure (recently measured or estimated from models)
14) Productivity index (PI, estimated or calculated)
15) FBHP, P_{wf} (recently measured or estimated from models)

Fig. 8-9 displays a high-level dataflow of the well test validation process.

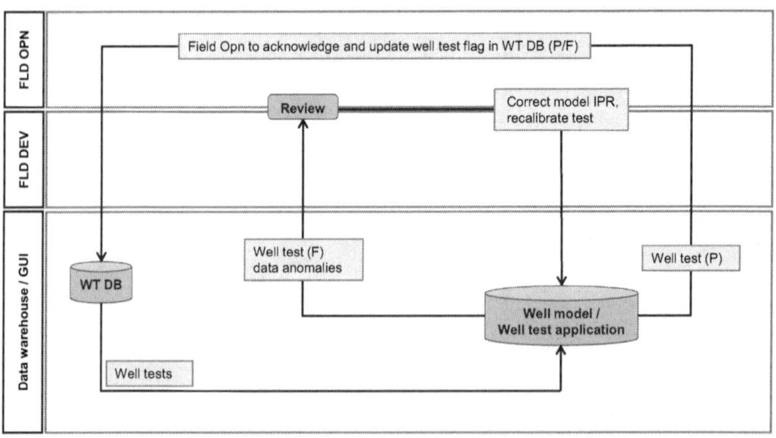

Fig. 8–9. Well test validation dataflow

The well test validation process follows the following sequence:

1) Well tests conducted through a test separator or portable well test unit or multiphase flow meter are stored in a well test database.
2) In an automated workflow, well tests for gas lifted wells are accessed and processed by a well test application. Measured values (values 1–9 mentioned above) from the well test are used to obtain the calculated values of well parameters (values 10–15 mentioned above).
3) The software program working with the inbuilt logic in the well test application compares the newly calculated well parameters with the historical performance, using elements of certain pre-set acceptance criteria. This will determine whether the well test will "pass" or "fail." The engineer-in-charge gets notification to this effect.
   - A passed well test (P) is sent directly to field operations to acknowledge and update the well test flag so it can be stored in the well test database and accessed for production allocation and other requirements.
   - A failed well test (F) is presented to engineers from field development and production operation for "review" and collaborative work on the reasons / actions suggested by the well test application.
4) The review will require work on data anomalies (in static and real-time data) and updating / recalibrating the well models with the latest IPR or well configuration information. It may also trigger the need to revise elements of the acceptance criteria for the well test.
5) A failed well test may be accepted as "pass" after the team's review or may be discarded permanently with a failed well test flag and a recommendation to re-conduct the well test.

Passed well tests will be used to generate new performance curves for the wells.

Fig. 8-10 shows how an ML model can be developed to assist in the well test validation process. The ML model developed can work as a substitute for the well test application or along with it, depending on the end-user requirements.

The following sequence will be followed when developing an ML model for the well test validation process:

1) Well tests are stored and managed by the well test database repository.
2) A data analytics algorithm is used for data mining (extract, transform, and load) of well test data available in structured and unstructured formats.
3) Well test data is split into two data sets: *training data set* (80%) and *testing / validation data set* (20%).
4) An ML model will be trained to learn the types of well tests, acceptance envelope of well test parameters, and characteristics of passed / failed well tests, based on the input well test parameters mentioned above.

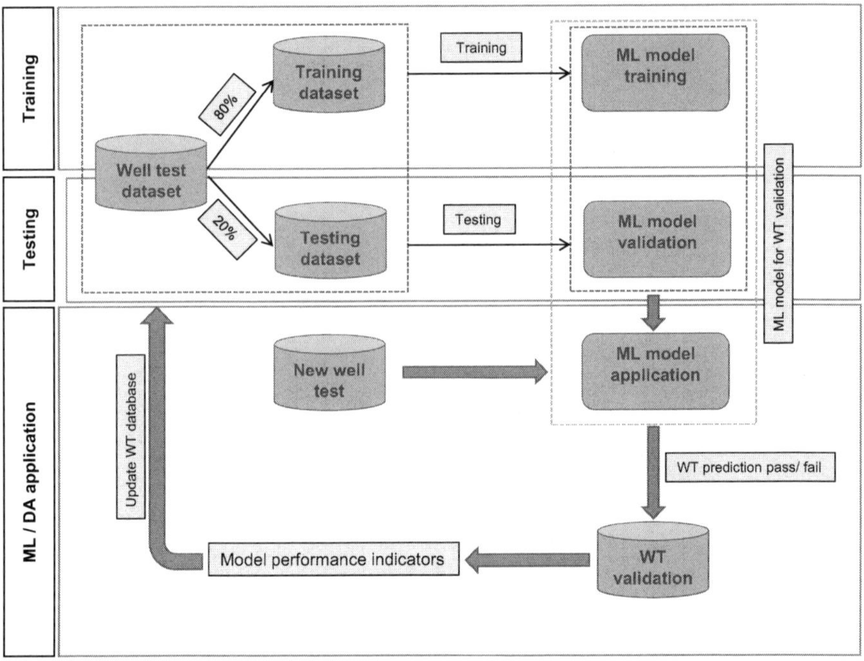

Fig. 8–10. Well test validation with DA / ML models

5) A single ML model or a composite ML model can be used to generate multilayered artificial neural networks to enable deep learning (DL), considering the span of input parameters and / or desired output.
   - For gas lifted wells, ML algorithms will be trained to learn and predict the possibility of shallow gas lift or multi-point lifting. This will be of significant value.
   - The opportunity for value generation by re-allocation of lift gas (in the event of a shortage of lift gas) may also be considered for the long term.
6) Once trained, the well test ML model will be validated against the remaining 20% of well test data. After achieving a level of accuracy, the well test ML model will be used as part of a bigger well test validation workflow. The ML model will continue to improvise and improve.

8.3.4 Freezing Problem

In the data plot (**Fig. 8-11**) of well A3 (referring to the data from **Table 8-1**), for the first six days, the well seems to be flowing normally. During days 7–8, the gas lift injection rate decreases and shows heavy fluctuation. This is accompanied by a decrease of the casing head pressure. The wellhead flowing pressure and temperature also register the change. The fluctuations initially start small with a freezing tendency. After 2–3 hours, the freezing really happens, and the parameters go down. The issue is identified immediately by a statistical analysis method.

The plot shows a bandwidth of 40–60% of the average value of gas lift injection rate in real time. The combination of events, i.e., the decrease and fluctuations in the rate and the decrease of casing pressure, are symptoms of freezing. When the gas lift injection goes below 80% of average, the freezing issue gets highlighted in the well. Later, injection of mono-ethylene-glycol as an antifreeze chemical is used to solve the freezing problem. The well becomes normal after two days.

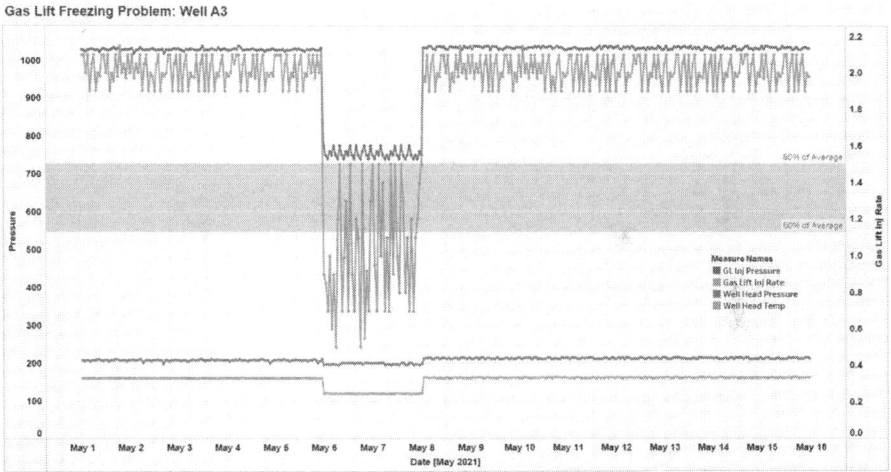

Fig. 8–11. Indication of freezing

8.3.5 Well Gas Lift Unloading Problem

Fig. 8-12 highlights the unloading problem of a gas lifted well (A5). The gas lift injection rate is well below the set point. The casing pressure is not steady and is very close to the gas lift manifold pressure. It affects the production rate of the well. The problem is identified when the gas injection rate remains lower than its defined threshold limit on a real-time basis.

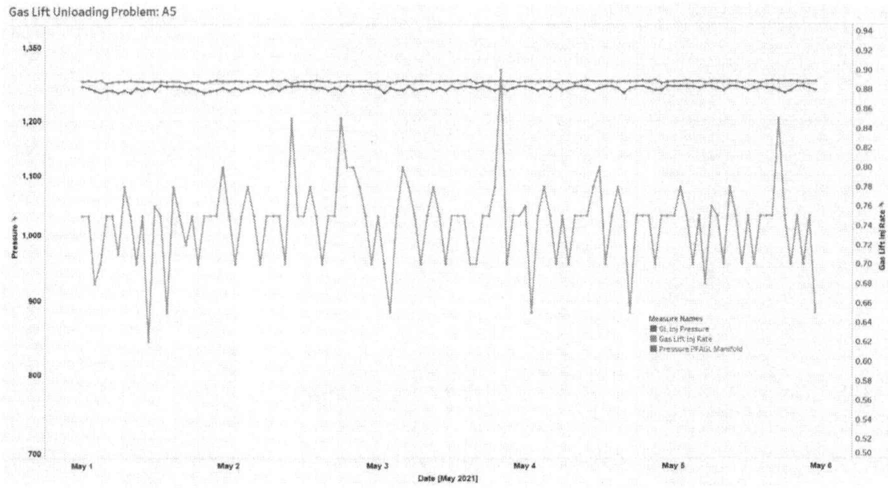

Fig. 8–12. Gas lift unloading problem

8.3.6 Well Performance from Flowline Temperature

In gas lifted wells, the oil flowline temperature gives a qualitative indication of the well flowing condition. The flowline temperature is affected by the well production rate, water cut, reservoir temperature, depth of the reservoir, GLR, and ambient conditions. The flowline temperature differs from one well to another. However, in any well, a change of flowline temperature could indicate one of the following:

- Increase in temperature, indicating a possible increase of water cut
- Decrease in temperature, suggesting well is flowing with lower water cut or at a lower rate
- Serious decrease in temperature, indicating lift gas circulation, not producing any fluid from formation
- Same as ambient temperature, suggesting no flow in the flowline (well is dead, shut-in, or SSSV is closed)

Fig. 8-13 shows the variation in flowline temperature in a gas lifted well. The flowline temperature initially decreases to a lower than normal value. Then it becomes almost equal to the lift gas supply temperature, indicating that the well is injecting lift gas through shallow mandrel(s) probably above the fluid level in the tubing.

When changes are happening in the gas lift injection pressure and flowline temperature, the automated gas lift choke is able to maintain the gas lift rate at the surface. Thus, no anomaly could be detected in the gas lift rate in this case.

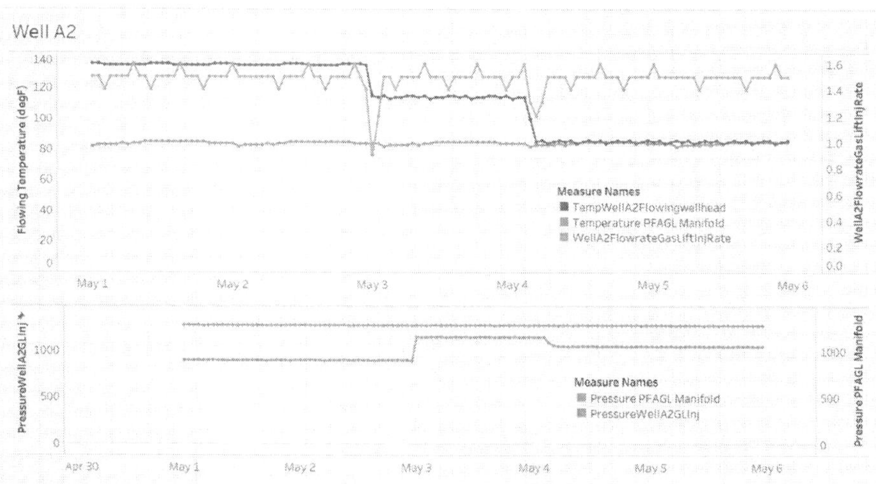

Fig. 8–13. Flowline temperature

8.4 Use of Hybrid Models for System Optimization

The question arises whether the physics-based models for the various components of the oilfield are still needed for system optimization or not. The answer is *yes*. The following example problems will clarify this point:

- Calculating the pressure or temperature drop in a pipeline that carries multi-phase fluids
- Estimating the output of a compressor or pump when an operating parameter is changed
- Predicting the response of a well when the production choke size is increased
- Performing sensitivity studies

Laws of physics are required to find possible answers to these problems. Even if these models are structurally complex, the fact that they are the proxy representations of a physical reality cannot be ignored. Therefore, the possible range of solutions that would come from the physical model should be utilized when generating optimization solutions.

Users of physics-based models might encounter a few practical problems: first, a full set of data may not be available on hand to fully describe the problem at a particular moment; second, due to their high level of complexity, the physics-based models may take a long time to arrive at solutions. But these two issues can be overcome.

Data itself cannot be an alternative for physical modeling, but when combined with informed and detailed knowledge of the physical problem and its constraints, it is likely to yield successful solutions. [42] Therefore, it would be prudent to pursue a balanced perspective.

As the data from these physical objects at different conditions represents the physical response of the actual equipment, the data from the field can be used to build an analytical model. The data-driven model so constructed can be trained using both simulation and field data.

If the available data points are insufficient to fully describe the physical model, the data-driven model can be designed to complement a physics-based model to increase computational efficiency. The hybrid models with improved predictive, prescriptive, and cognitive capabilities will give better solutions at a better pace. This will provide the benefits of both systems.

A highly complex, physics-based model can thus be simplified without losing its value, with the help of a data-driven model for optimization. The workflow is shown in **Fig. 8-14**.

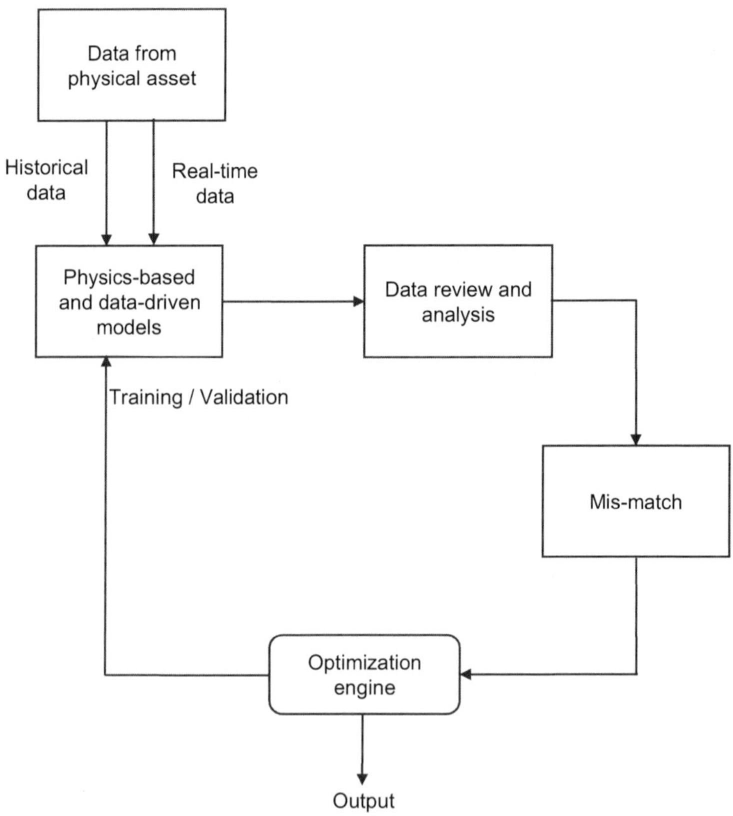

Fig. 8–14. Use of hybrid model for optimization

9

Digital Integration of Assets

The "producing" and "consuming" units of an asset are functionally linked in the production, movement, and processing of hydrocarbons. The digital data integration of these assets can do wonders. Precisely, the Real-time Operations Center (Production) aims to integrate the operations and automate the workflows to achieve a higher level of efficiency for all the divisions or companies connected with the asset. With data reliability and transparency, the future opens up for further integration, achieving synergy among various companies engaged in the energy sector.

The following topics are discussed in this chapter:

- The Field Operations Center (FOC)
- The Real-time Operations Center (Production)—RTOC (P)
- The interdependencies of upstream, midstream, and downstream operations
- Expanding the horizon for data integration outside the organization

9.1 Data Integration across Assets

The data-sharing that is practiced within an asset can be expanded to the whole company, then to the region, and across the country. The producing and receiving units can be brought under one umbrella of automated operations so that efficiencies are fully achieved. Companies involved in the hydrocarbon chain—irrespective of being owned by government, semi-government, or private—can collaborate with each other to achieve the benefits of synchronized operations. Establishing data integration across various operating units helps achieve the benefits of the collaboration.

9.1.1 Integrated Oil and Gas Operations

Oil and gas operations are categorized into upstream, midstream, and downstream activities. The upstream activity involves oil and gas exploration and production, the midstream activity focuses on storage, processing, and transportation, and the downstream activity deals with petroleum refinement and marketing. In a country that has reasonable quantities of hydrocarbon reserves, it would be beneficial to integrate oil and gas operations by effectively linking all these activities. This may be done in two ways:

- **Through integrated companies**

  Integrated companies have individual business divisions dedicated to the upstream, midstream, and downstream sectors of the oil and gas industry. Their assets consist of equipment used in exploration and drilling, transportation infrastructure (trucks, tankers, or pipelines), refineries, and even gas stations. They are companies such as Chevron, ExxonMobil, and BP. An integrated company is able to have complete control of its operations and achieve improved efficiency. It is able to bring in various streams of revenue and diversification. However, some of these companies have split their upstream and downstream operations due to pricing issues of the raw materials and finished products.

- **By integrating operations**

  A government authority coordinates with individual companies that are involved in the entire value chain of the oil and gas industry. The role is to identify synergies in operations with the aim of achieving a higher level of productivity and efficiency across assets. In the absence of this coordination, companies tend to stick to their individual goals and do not readily respond to the changes that are happening outside their operating environment. The integration process would ensure inter-asset collaboration to the extent possible.

Fig. 9-1 illustrates the product flow across various operating units of a multidisciplinary oil and gas production company that operates four assets. The interdependencies among various operating units dealing with production, transportation, and consumption or sales of hydrocarbons will be demonstrated.

- *Asset A* mainly produces oil and associated gas. It would need lean gas for gas lifting wells and for gas injection into the reservoir.
- *Asset B* mainly produces gas. The produced gas would be sent for gas processing. But it also requires lean gas for injection into the reservoir.
- *Asset C* has both oil and gas production wells. It would need lean gas for gas lifting wells and for gas injection into the reservoir.
- *Asset D* is functionally the same as Asset A, but quantities are different.

The following points describe how these assets and export units influence each other:

1) Every oil-producing asset has its own daily, monthly, and quarterly targets and a production quota that should be met based on the overall commitments from shareholders. Combined oil production from all assets gets bifurcated to (a) export, based on monthly tanker schedule and (b) refinery, for domestic consumption and export of oil products.

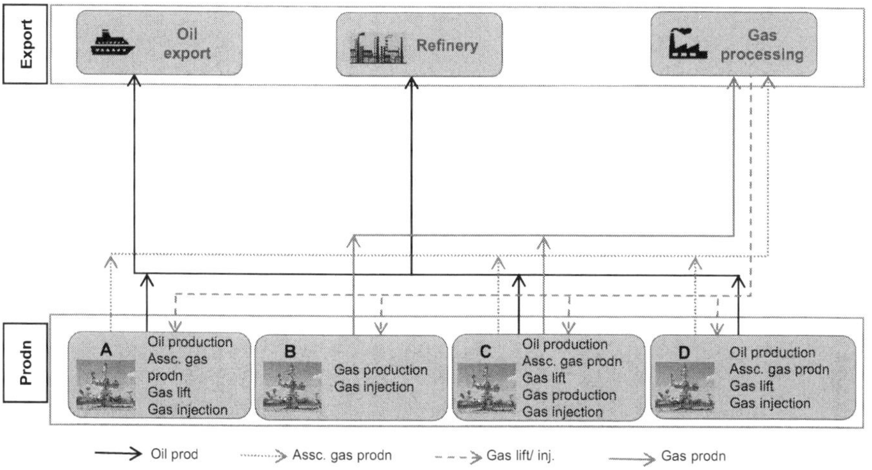

Fig. 9–1. Interdependencies of assets

2) An oil-producing asset cannot produce more than allowable technical rates defined by the reservoir team. These rates have upper limits in order to maintain the health of the reservoir, i.e., maximize ultimate recovery of hydrocarbons.
3) Every gas-producing asset has its own daily, monthly, and quarterly targets and a production quota in order to meet the overall gas production quota from shareholders for gas sales and domestic use.
4) Gas assets also cannot produce more than allowable technical rates for the gas reservoir.
5) The associated gas produced from oil wells becomes a part of gas production. The gas processing unit should return lean gas for gas lifting the wells and providing gas injection into reservoir. From a gas balance perspective, the lift gas is returned back into the production system, but gas injected into the reservoir is removed from the surface loop.
6) If gas production exceeds the capacity of the gas-processing units, then either gas production or associated gas production should be reduced. If associated gas is reduced, this, in turn, reduces oil production, and this upsets the asset's oil production target.
7) In the event of oil production upset due to unplanned / planned shutdown in any asset, there will be a need to ramp up production from other asset(s) to maintain the total oil export target and tanker lifting schedule. The reservoir production guidelines should still be followed.
8) In the event of an export upset (inclement weather, issues related to oil storage, etc.), there will be a need to curtail oil production from

oil-producing asset(s), but this would require ramping up gas production from gas assets so that the gas sales targets could still be met. If there is an option to send oil to an alternate storage terminal, the oil production from the field need not be reduced.

It is clear that each operating unit (OU) aims to achieve its own production goal and KPIs, but at the same time should make a productive contribution toward managing the interdependencies of other OUs / sister companies in order to meet the energy requirements of the nation.

For example, if there is a production disruption in field B, the gas export is affected. This would immediately affect the operation of the gas-processing plant. That would, in turn, affect the lift gas supply to assets A, C, and D. The chain reaction would impact the oil production from the three assets. If not rectified immediately, the impact will cascade and eventually affect the operation of the oil storage and loading terminals, as it may not be possible to load oil to the tanker as per schedule.

By linking the real-time data of all the production assets and the export units, it becomes possible to make necessary adjustments in the connected units with the aim of streamlining the operations.

9.1.2 Field Operations Center

The individual assets A, B, C, and D described in the previous section cannot be simply treated as "supply tanks" that can supply infinite quantities of oil and gas. The reservoirs, wells, pipelines, processing vessels, and machineries in each field have operating envelopes and design limitations. Due to aging, they may have additional operational challenges too. When extracting hydrocarbons from these assets, these operating envelopes should be honored, based on the characteristics of reservoir, completion, and process equipment. This would ensure proper production management and optimization of each asset.

With real-time optimization established, the oil and gas production and injection portfolios of each asset will be optimized with the help of a *Field Operations Center*, also called *Collaborative Working Environment*. Each asset has an FOC. These FOCs utilize the real-time instrumentation and sensor data from wellheads, pipelines, and process plants combined with modern technologies to optimize production and injection, maintaining efficiency at top levels. The FOCs are equipped with tools and strategies that would enable assets to meet their production targets, corporate, technical, and various other KPIs. In other words, FOCs assist in the operational management of the entire asset, paying attention to each and every surface and sub-surface component of the asset.

9.1.3 Real-time Operations Center (Production)

Even though optimization of each asset is important, equally important for the assets is meeting corporate targets. These targets are assigned by governmental agencies or partners. The Real-time Operations Center (Production), through its automated workflows, helps assets achieve both the targets by facilitating inter-asset optimization.

The concept of RTOC (P) evolves from the Real-time Operations Center (Drilling) conceptualized in the 1970s. RTOC (D) helps in enabling complex wells to achieve drilling and production targets with significant cost savings. Such a concept applied to production operations is called RTOC (P). Here the value addition is ensured by connecting data from all the assets so that the assets work in unison, helping each other when needed to meet the corporate goals, while striving to maximize the efficiency of each asset. This should be done on an ongoing basis by sharing and using live data.

In the present era, where digital innovation is catalyzing economic progress in the oil and gas industry, the RTOC (P) makes perfect sense. In fact, unplanned downtime of equipment in one asset can become an opportunity to try a different mode of operation to compensate for the loss of production—or to carry out a pending maintenance job in another asset, making use of the opportunity.

The RTOC (P) gets asset performance data and the status of critical equipment through the FOCs. It also has access to the outcomes of various analyses from several asset and equipment models available in the FOC. So the RTOC (P) knows what an asset can deliver, if it demands more from the asset, considering the current status and the prescriptive analysis of various systems of the asset, evaluated through neural networks and artificial intelligence. On top of this, special algorithms for maximizing multi-asset production are incorporated within the RTOC (P) to facilitate inter-asset optimization. The optimization can be in manual, semi-automated, or fully automated modes.

The RTOC (P) will be able to interact with all the producing assets and delivery points such as oil delivery, gas processing, and refinery. Should some unplanned event happen in one asset, through the automated workflows, it would swing into action to manage the process upset in the most profitable manner. The goal would be to have minimal disturbance to the upstream, midstream, and downstream operations under its purview and to minimize the overall loss of revenue to the extent possible.

RTOC (P) fosters better interaction and more productive collaboration at all levels within a company. The same concept can be expanded to several organizations in the hydrocarbon chain within a country.

9.1.4 Functional View of the RTOC (P)

Just like the process connectivity existing in the field, the dataflow should be established among various physical entities involved in the hydrocarbon chain. **Fig. 9-2** shows the dataflow between the producing assets and the receiving units, all communicating back and forth through the RTOC (P). The zone of influence of the RTOC (P) is highlighted.

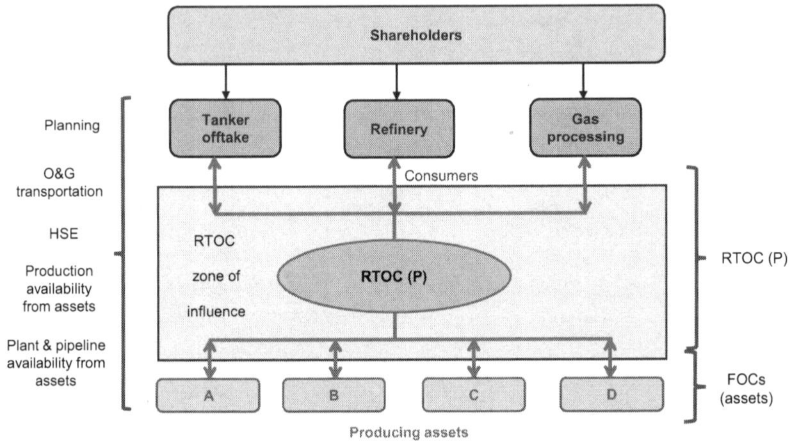

Fig. 9–2. RTOC (P) dataflow and zone of influence

The RTOC (P) has emerged as a new age tool for corporate management to strike an optimum balance between the technical limits of the assets and commercial deliverables. All the physical and analytical models that have been set up at the asset level can be used by the RTOC (P). In short, it makes use of all the technical work done so far in optimizing the assets.

9.1.5 Physical Setup of the RTOC (P)

An FOC serves as a workstation for the members of an asset. Data from the field for real-time surveillance and optimization is displayed in a conference room within the same building where the asset team members work. This room has the facility to display the field parameters on a large display. This room also has an AV (audio-video) conference facility for engineering, field operations, and technical office staff to connect to each other or with any other vendor location, when needed. This will be used by the members of the asset team to manage the day-to-day operations of the asset. In short, the FOC integrates various disciplines of the work force within an asset.

Like the FOC, the RTOC (P) is physically established in the main office building at a central location. The RTOC (P) has the facility to drill down any asset to show the full view of the dashboard of the FOC of that field, subject to login privileges.

One can even drill down further to know the condition of an individual object like a pump, compressor, or a well in any of the fields, when required. Login privileges for individuals will be decided by the organization—some may be able to access both the FOC and the RTOC (P), whereas some others will be able to access the FOC only.

The RTOC (P) integrates the assets, terminals, gas processing plant, and refineries that are linked to the assets of the organization. The customized workflows to be executed at the corporate level would be built into the RTOC (P). The launch page of the RTOC (P) can display high-level results and dashboards. Along with the corporate production targets and KPIs, the updated results can be displayed using various visual presentation techniques.

Functionally, the RTOC (P) can also be integrated with the RTOC (D) and Incident Command Center (ICC). The ICC is a standby setup in a company that becomes active if there is any major incident or accident that can disrupt the operations.

It is also possible to launch the RTOC in a virtual mode. The setup of the RTOC (P) in a physical building may sound like a requirement to get the necessary visualization to see all the data together and be able to work together. However, this can be done from any conference room of the company that has the basic IT setup and screens. A management team member with the required privileges can even open up the RTOC (P) screen from any PC connected to the network.

9.1.6 Technology Used by the RTOC (P)

Arming staff with the latest technology and connectivity to all areas of operations, the RTOC (P) is able to provide the best solution in the automated online mode, as well as for stand-alone decision-making. The RTOC (P) utilizes the latest technologies that are available. The tools utilized by the RTOC are displayed in **Fig. 9-3**.

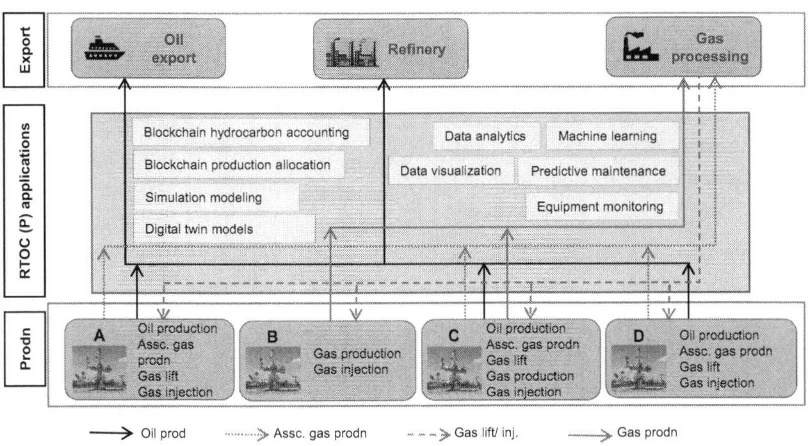

Fig. 9–3. RTOC (P) deployment and functioning

Some of the technologies that belong to *Industrial Revolution 4.0* that are currently used in RTOC (P) worldwide are briefly explained below.

- **Prescriptive Maintenance**:

  In an environment of the IIoT, RTOC (P) makes use of prescriptive maintenance generated in the FOCs by the use of various data analytics, machine learning, and artificial intelligence models. The aim is to improve the reliability of various rotating and process equipment items in the asset and avoid unplanned shutdowns.

- **Simulation Modeling**

  RTOC (P) makes use of *simulation modeling* in an extensive manner. The integrated asset models for reservoir, wells, pipeline, and facilities already created through engineering applications provide optimized results for every asset. These results can be used as inputs to the inter-asset optimization algorithms that are handled within the RTOC (P). This will increase the value of every barrel of oil produced by the assets.

- **Digital Twin Models**

  A *digital twin model* is a replica or virtual representation of a real-life object, process, or system. The idea is to exactly simulate the physical to virtual, with the capability of handling real-time instrumentation and sensor data from the real world. These digital twin models are equipped with data analytics and machine learning algorithms that can create various *what-if* scenarios. The RTOC has access to such digital twin models of various process systems across the assets it is connected to. The twin models will also shorten the cycle of decision-making for project selection by providing enhanced technical and economic insights.

- **Blockchain Technology**

  Blockchain is a system of recording data or information in a chain of blocks in such a way that it cannot be changed, manipulated, or hacked. It works as a digital ledger of transactions that is duplicated and distributed across computer systems on blockchain. This emerging technology has many applications in oil and gas, especially for production allocation and reconciliation between assets. This would be extremely useful for hydrocarbon accounting by enabling smart contracts between producer, buyer, seller, and consumers, all having *one* version of the data.

9.1.7 Visual of the RTOC

Fig. 9-4 shows the screen display of an established real-time operations center in London (the designer called it a "Unified Operations Center"). The primary focus of the RTOC in this case is connecting people, assets, and data across the organization.

Fig. 9–4. Real-time operations center screen display
Image courtesy: AVEVA [43]

Apart from the main screen display, there will be additional screens and cubicles for people to review specific areas of the asset with live data and to conduct virtual meetings with subject matter experts (SMEs) to decide on maintenance strategies and equipment troubleshooting.

9.2 Stepping into the Future

9.2.1 RTOC Alerting Corporate Target Shortfalls

A company manages several targets on a daily, weekly, and monthly basis. Some of the key information the management may be interested in for every asset includes the following:

- Production of oil and gas
- Injection of gas and water
- Health, safety, and environmental (HSE) performance
- Carbon dioxide (CO_2) emission
- Sub-surface capacity available
- Well availability
- Plant availability
- Critical equipment availability
- Several KPIs established already

The RTOC (P) is customized to retrieve and display the full information on what the corporation has to deliver by the end of the day, month, or year, as well as the present status. This will guide all personnel to work together and focus on actions to meet the corporate goals.

What the FOC does for a single asset is done by the RTOC for the whole company that is operating several assets.

A simple production tracking function of the FOC will be demonstrated for an asset. The asset has three fields: X, Y and Z. The screen display as seen on the 24th day of a month is reproduced. The target vs. actual daily production rates in STB Oil/Day for the three fields and the total asset are shown in **Fig. 9-5**.

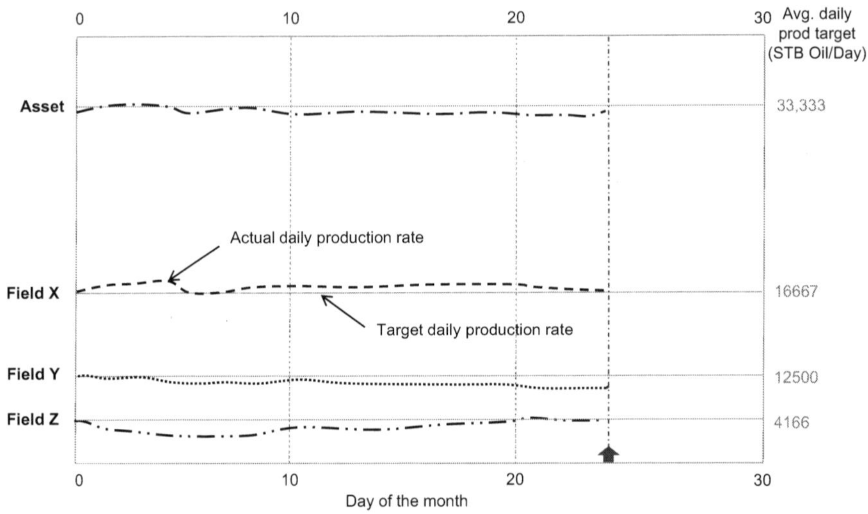

Fig. 9–5. Daily production rate: target vs. actual

It is clear from this plot that the asset production is slightly below target. Field X seems to be producing more than its target, but fields Y and Z seem to be below their targets.

The exact excess or shortfall is known from the cumulative plot for the month till date, shown in **Fig. 9-6**.

As can be seen, the actual production from the asset is slightly lower than target (yellow flag). The cumulative production is lower than the target by 25,000 STB oil (3.1%, in this case). Field X with 50% contribution for the asset has contributed more than its target till this time (*green*); however, Fields Y (37.5% contribution) and Z (12.5% contribution) are behind targets (both *red*). So during the remaining days of the month, production should be increased to meet the asset's target for the month. The displays are helpful to alert the fields to initiate necessary action to meet the production target of the asset.

With a similar approach, the RTOC can display the production trend and its shortfall from various assets.

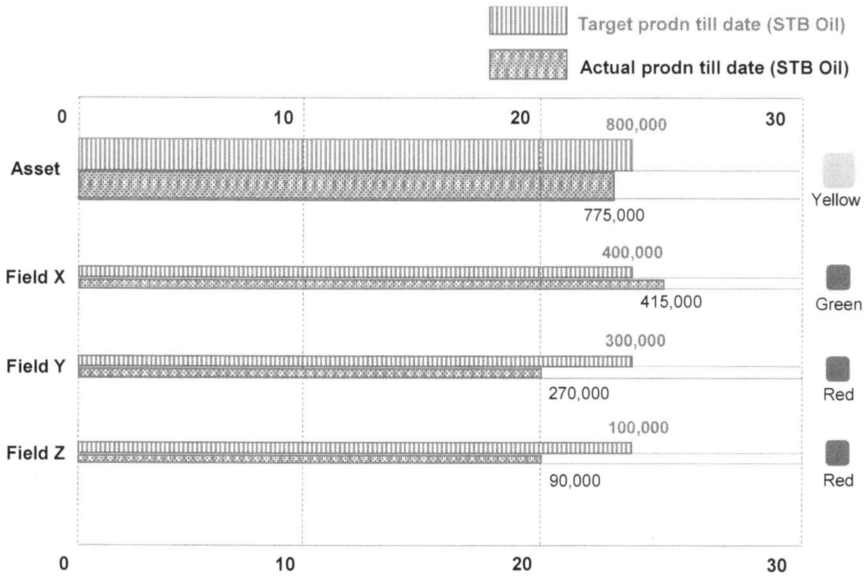

Fig. 9–6. Cumulative monthly production: target vs. actual

9.2.2 Working beyond Borders

The concept of data integration can be extended to any level. It can be extended among several companies in a state—or across several states or regions of a country. It can also be established to manage resources in a consortium of several countries. In such a situation, the system will work and yield superior results if:

- The assets are physically connected through inter-state or cross-country pipelines
- Business objectives of the integrated operations are clearly defined
- The current contribution and limit of each asset or operating unit are correctly factored in

If the regular oil storage is full, the RTOC (P) will find a nearby oil storage that belongs to another operator and divert the oil production to the alternate location for the next two to three days, so that production operations need not be shut down in an asset. If all the gas lift compressors are down, it will find a nearby source of high pressure gas that is technically acceptable in terms of pressure and composition that can be re-routed on a temporary basis to continue gas lift operations in the gas lifted asset, so that the production need not be shut in. The possibilities are endless.

9.2.3 Center of Excellence

Going beyond production management, the RTOC (P) can facilitate the sharing of collective knowledge gained across assets with all personnel linked through the operation chain.

Some of the most advanced companies have adopted centers of excellence where engineering staff are able to bring together engineering knowledge from failure analysis methods, such as root cause failure analysis (RCFA), in order to address recurring problems. Centers of excellence open up the horizon for knowledge-sharing and seeking solutions for field problems. The problems encountered or lessons learned at one location can be shared across the entire interest group through the RTOC (P). Contribution from multiple assets would support decision-making and maintenance planning, and this may even assist in future equipment design modifications.

In the old culture, most of the time the field would be fighting fires with their own equipment failures. The only recourse available was the equipment supplier, but sometimes that did not work efficiently due to the absence of a win-win spirit. With the adaptation of the new culture, field management can focus on sharing valuable information across operating entities, service company engineers, and SMEs.

The RTOC enables many traditional oilfield roles to be performed from the safety of an office environment, thus eliminating or reducing travel time and hotel stays. Additionally, key engineering and discipline experts can be called on, as needed, without having to potentially spend days or even weeks on location, thereby increasing overall productivity. As it is possible to monitor operations remotely, the need for experts to visit field locations is decreased. The RTOC has given the industry the ability to reduce the number of field personnel without affecting the quality of operations. This is a big achievement in terms of HSE.

If a problem develops in a gas lifted well or the network, the data analytics tools deployed in the asset would propose a way forward. Field personnel may be able to spot some of the recurring problems and may have solutions based on experience. Even after implementing the known remedial solutions, it would be value-adding to establish a course of action to avoid recurrence of similar problems in other areas of the same asset or in other assets. This may require the services of a specialist. The expert may be sitting on the other side of the globe. With the FOC and RTOC (P) operational, it is not difficult to share old data and live data from the operations and get the opinion of specialists.

The same approach will work with any problem related to upstream, midstream, and downstream operations, once it is brought under the influence of the RTOC (P).

9.2.4 The Sky is the Limit

In summary, the oilfield is poised to gain significant benefits through model-based real-time optimization, combined with the application of analytical tools from big data analytics. The FOC will make multiple disciplines of an asset work as a single force, whereas the RTOC will make several assets work as a single unit. This will improve vertical and horizontal communication. The sharing of information will improve accountability among individual team members and promote teamwork of a high order.

The RTOC opens up the doors of constructive collaboration with external agencies at a level that could not be imagined in the past. All this is now possible because of data transparency, that is, a single version of data seen by all parties in real time. When needed, the system will make it possible to seek help from subject matter experts almost instantaneously, irrespective of their geographical location. Operating companies do not have to wait days or months for a recommendation from an expert. It will all become available then and there. The members of the oil and gas community should prepare themselves to become a part of this new world of infinite possibilities.

Appendix

A1

List of References

1. James F. Lea; Henry V. Nickens, *Selection of Artificial Lift,* Paper presented at the SPE Mid-Continent Operations Symposium, Oklahoma City, Oklahoma, March 1999. Paper number: SPE-52157-MS.

2. Brown, K. E., et al.: *The Technology of Artificial Lift Methods – Volume 2a,* PennWell Publishing Co., Tulsa, Oklahoma, 1980.

3. ReO™ Instruction Manual and Product brochure on Real-time optimization (Weatherford Production Optimization).

4. Kashif Rashid, William Bailey and Benoit Couet, *A Survey of Methods of Gas-Lift Optimization,* Hindawi Publishing Corporation, Modelling and Simulation in Engineering, Volume 2012, Article ID 516807.

5. Extracted from Encyclopedia Article from https://www.britannica.com/topic/simplex-method, article revised and updated by William L. Hosch, Associate Editor.

6. Handley-Schachler, S., McKie, C.J.N., et al., *"New Mathematical Techniques for the Optimisation of Oil & Gas Production Systems",* paper SPE 65161, presented at the SPE European Petroleum Conference, Paris (2000).

7. GAP Instruction Manual (Petroleum Experts Ltd.)

8. Stewart, G., Clark, A.C., and McBride, S.A., *"Field Wide Production Optimisation",* paper SPE 59459, presented at the SPE Asia Pacific Conference on Integrated Modelling for Asset Management", Yokohama (2000).

9. M. S. Nadar, R. Kulkarni, Z. R. Lemanczyk, and P. Aggarwal, SPE 101089: *"Development of an Integrated Production Network Model for the Heera Field";* [2006 SPE Asia Pacific Oil & Gas Conference and Exhibition held in Adelaide, Australia, 11–13 September 2006].

10. Manickam S. Nadar, Tim S. Schneider, Kathy L. Jackson, Calum J N. McKie and Javad Hamid, SPE 103670: *"Implementation of a Total System Production Optimization Model in a Complex Gas-lifted Offshore Operation* [2006 SPE Annual Technical Conference & Exhibition (ATCE) held in San Antonio, Texas, USA, 24-27 September 2006]. *This paper was selected and later published in the SPE Production & Operations Journal.* [SPE Production and Operations, vol. 23, no. 1, pp. 5–13, 2008].

11. Bin Amro, A.A.; Sakria, D.; Lestariono, Y.; McAlonan, N.; Shere, A.J. *"Benefits From Implementing an Integrated Asset Operations Modelling System"* Paper SPE

127893 presented at the SPE Intelligent Energy and Exhibition held in Utrecht, The Netherlands, 23-25 March 2010.

12. Jha, Anand Kumar; Barua, Santanu; Agrawal, Vikas; Agrawal, Gunja; Kumar, Ravi; Moitra, S.K.; Negi, G.S. *"Implementation of Integrated Network Optimization Model for the Mumbai High Field-Crucial to Field-Wide Optimization"* Paper SPE 123799 presented at 2009 SPE offshore Europe Oil & Gas Conference & Exhibition held in Aberdeen, UK, 8-11 September 2009.

13. Kumar, Aditya; Omer, Omer Mohamed; Sulaiman, Abu Baker Ahmed; *"A field Wide Surface Network Modeling and De-bottlenecking of Production Network for Nine Fields of GNPOC"*. Paper SPE 150555 presented at North Africa Technical Conference and Exhibition held in Cairo, Egypt, 20-22 February 2012.

14. Fathi Shnaib, Manickam S. Nadar, Steve Booth and Ademola Otubaga, SPE 126681: *"ESP Application to De-bottleneck Gas Lift System Offshore"*; Paper presented at the 2010 SPE Oil and Gas India Conference and Exhibition held in Mumbai, India, 20-22 January 2010.

15. Dr Z. Rick Lemanczyk, Production Optimization Project Notes, Weatherford Production Optimization.

16. Dr Rajan Chokshi, Gas Lift Training Notes, eProduction Solutions, Weatherford Production Optimization.

17. Crane Technical Paper No. 410 (TP-410).

18. Nadar, M.S., Technical paper on *"Benefits of Detailed Compressor Modeling in Optimizing Production from Gas-lifted Fields"* in the Middle East Artificial Lift Forum (MEALF), June 2004. This was later published in the PIPELINE magazine.

19. Frank J. Brooks, "GE Gas Turbine Performance Characteristics" (GER-3567H), Paper published by GE Power Systems.

20. ReO Forecast literature, EPS Ltd., Weatherford Production Optimization.

21. Jane Whaley *"What is a Digital Oilfield?"* article published in GeoExPro (The favorite petroleum geoscience magazine), Vol 12, No.2 – 2015.

22. PetroWiki, *Intelligent wells: Summary on Intelligent Wells.*

23. Salem Aldousary, Extracts from paper on *Voidage replacement ratio (VRR)*, submitted as coursework for PH240, Stanford University, Fall 2015, December 5, 2015.

24. Hong-Quan (Holden) Zhang (Director of the University of Tulsa Artificial Lift Projects and an associate professor of petroleum engineering), *Artificial Lift Optimization: Horizontal Well Characteristics Impact Artificial Lift Performance In Unconventional Reservoirs*, American Oil and Gas Reporter, June 2015 Cover Story.

25. Trent Jacobs, *Life After 5: How Tight-Oil Wells Grow Old*, Journal of Petroleum Technology, January 31, 2020.

26. Pankaj, P., Patron, K. E., and Lu, H. *Artificial Lift Selection and Its Applications for Deep Horizontal Wells in Unconventional Reservoirs*. Paper presented at

the SPE/AAPG/SEG Unconventional Resources Technology Conference, Houston, Texas, USA, 23–25 July, 2018 URTEC-2875180-MS. https://doi.org/10.15530/URTEC-2018-2875180.

27. Jay Miller, Production Lift Companies Inc., "*A life of the well artificial lift strategy for unconventional wells*", presented at Artificial Lift and Production Congress, January 29, 2019.

28. Sutton, R.: "*Wellbore Geometry Effects on Well Production Performance*," SPE Liquids Rich Shale Conference, Rancho Palos Verdes, CA (May 2013).

29. William Lane, *Completion and Artificial Lift Strategies for the Life of the Well*, Weatherford presentation for ALRDC.

30. Darryl Polasek and Chandler Frost, *Extending the Range of Gas Lift to Long Perforations*, Paper presented at the SPE Annual Technical Conference and Exhibition, San Antonio, Texas, USA, October 2012. Paper Number: SPE-160003-MS.

31. Kenneth Estrada, Concho Resources Inc., and Jay Miller, Production Lift Companies, "*Alternative Gas-lift Design For Unconventional Wells: Make more fluid with less gas.*"

32. M. A. Usmanl, A. O Olatunde, T. A. Adeosun, and O. L. Egwuenu, *Hydrate management strategies in subsea oil and gas flowlines at shut-in condition*, Department of Chemical Engineering, University of Lagos, Nigeria, Department of Mathematics, Yaba College of Technology, Nigeria, Received January 30, 2012, Accepted August 15, 2012, ISSN 1337-7027.

33. Subash S. Jayawardena, George J. Zabaras, Leonid A. Dykhno, *The Use of Subsea Gas-Lift in Deepwater Applications*, Paper presented at the Offshore Technology Conference, Houston, Texas, U.S.A., April 2007, Paper Number: OTC-18820-MS.

34. Brule, M. R., "*Big Data in E & P: Real-time Adaptive analytics and dataflow architecture*". Paper SPE 163721 (2013).

35. Bravo C. et al. *Applying Analytics to Production Workflows: Transforming Integrated operations into Intelligent Operations*, Paper SPE 167823 – MS (April 2014).

36. "What is data Analytics?" Web article https://www.mastersindatascience.org/learning/what-is-data-analytics/ (powered by 2U, Inc.), Updated February 2021

37. Summary on Machine Learning from Expert.ai Team (May 2020).

38. Jason Brownlee, *Basic concepts in Machine Learning* in Start Machine Learning, December 25, 2015.

39. Jake Frankenfield, reviewed by Gordon Scott, extracts from article on Artificial Intelligence (AI), (March 2021).

40. Extracted from webinar: https://www.kdnuggets.com/2015/12/how-do-neural-networks-learn.html

41. Web article on IIoT: https://www.i-scoop.eu/internet-of-things-guide/industrial-internet-things-iiot-saving-costs-innovation/industrial-internet-things-iiot/

42. Hector Klie, *A Tale of Two Approaches: Physics-Based vs. Data-Driven Models: The Way Ahead*, published May 3, 2021, https://jpt.spe.org/twa/a-tale-of-two-approaches-physics-based-vs-data-driven-models
43. AVEVA Unified Operations Centre, Picture courtesy: https://www.aveva.com/en/solutions/operations/unified-operations-center/

A2

Data Required for Network Modeling

1.0 Fluid Properties

Depending on the optimization requirements, fluid properties are modeled using black oil or compositional mode.

1.1 Black Oil Fluid Data

For each fluid type to be represented in the model:

- Produced oil/gas gravity
- Condensate gas ratio (CGR)
- Condensate gravity if CGR > 0
- Concentration of inorganics: N_2, CO_2, H_2S
- WGR and water density

1.2 Compositional Fluid Data

For each fluid type to be represented in the model:

- Well stream composition of produced gas, including inorganics (e.g., N_2, CO_2, H_2S) and hydrocarbon components to at least C7+
- Molecular weight and density of hydrocarbon plus-fraction (Cn+).

For compositional fluid modeling, the system modeling software uses a fully compositional equation of state (EOS) model. The predictions of the EOS model can be tuned to match observed laboratory PVT data if this is provided. Several fluid compositions can be entered to represent varying fluid properties from different parts of the field(s) being modeled.

It is also possible to enter fluid composition information in terms of black oil data (oil/gas gravity, condensate gravity, CGR, WGR), which will then be mapped to a composition. It is possible to enter any combination of compositional and black oil models, but at least one "base" composition must be entered.

2.0 System Geometry

The configuration and connectivity of the surface production network will be represented in the model in sufficient detail to meet study objectives. The relationship between the different items of equipment and the connecting pipework will be entered into the model by constructing a process flow diagram using the graphical user interface (GUI).

2.1 Network Topology

Diagrams are required, giving the relative locations and connectivity of surface equipment of the following types, where relevant to solving the problem:

- Wellheads
- Pipelines/flowlines
- Manifolds
- Compressors
- Heat exchangers
- Separation vessels (e.g., condensate recovery, liquid knockout, slug catchers)
- Surface chokes
- Block (isolation) valves
- Pressure control valves (PCVs)
- Flow control valves (FCVs)
- Delivery/sales points

Likely sources of information for the data required for this section are process flow diagrams (PFDs), piping and instrumentation diagrams (P&IDs), isometric diagrams, and any other schematics of the field that are used by operations and maintenance personnel.

2.2 Flowlines and Pipelines

For each flowline or pipeline, the following information is required:

- Length
- Outside diameter (OD)
- Internal diameter (ID)
- Elevations of each pipe end or angle of inclination from horizontal
- Pipe wall material (e.g., carbon steel, stainless steel)
- Thickness and thermal conductivity of any insulation materials
- Surrounding media (air, water, earth) and depth of burial, if appropriate

3.0 Wellhead Performance Tables

Wells will be modeled in the system modeling software using fluid source objects that represent the wellheads and contain a table giving production rate and wellhead temperature as a function of wellhead pressure. The data in the tables can **either** be entered directly **or** imported automatically from a flow model of the well.

3.1 Direct Entry of Wellhead Performance Tables

The following data is required if the wellhead performance data is to be entered directly. If the data is provided in this way, there is no need to provide well flow models (following section).

- A wellhead performance table is required for each well to be modeled. This consists of separator oil/gas rate and wellhead temperature as a function of wellhead pressure. Ideally, the range of wellhead pressures in the table should span the entire region between the shut-in well head pressure (at zero production rate) and 14.7 psia. If the entire range of pressures is not provided, extrapolation will be performed.
- The well's CGR and WGR as a function of wellhead pressure should also be provided if they are significantly greater than zero. The process (separator) conditions associated with the CGR and WGR values should be the same as those associated with the values of the separator gas rate provided for the previous item. In this way, the recombined well-stream molar flow rate will be calculated correctly.
  - Note that the values of CGR and WGR might not vary with wellhead pressure for any given well. However, the facility to model varying CGR and WGR with drawdown is available, if required.
- Where more than one fluid type is provided in the Fluid Properties section, a table is required identifying which fluid type each well produces. Note that if the fluid type assigned to a well was provided using black oil data, there is no requirement for the CGR in the fluid model to be the same as the value(s) of CGR provided in the wellhead performance tables. The composition of the well-stream will be automatically adjusted to be consistent with the CGR declared in the wellhead performance tables.

3.2 Import of Data from Well Flow Models

The following data is required if the wellhead performance data is to be entered from well flow models. If the data is provided in this way, there is no need to provide wellhead performance tables (previous section).

4.0 Compressor Data

Compressors will be modeled in the network using either the simple model or the detailed model. The simple model is suitable for all compressor configurations and models the compressor in terms of gas throughput, power, and compression ratio. The detailed model is only suitable for centrifugal compressors driven by gas turbines and incorporates the effect of compressor and turbine speed on performance. Surge and stonewall limits and recycling of compressor stages to avoid the former are also modeled.

4.1 Simple Compressor Model

The following data is required for all compressor trains that will be modeled using the simple model.

- Number of stages in compressor train
- Location of inter/after-coolers

- Maximum power available to compressor train
- Maximum compression ratio across train and/or individual stages
- Compressor train or stage mechanical efficiency factor
- Polytropic efficiency to be assumed for each compressor stage

5.0 Other Equipment

5.1 Heat Exchangers

The following data is required for each heat exchanger to be modeled:

- Heat exchanger service fluid (e.g., air, water, process-process)
- Heat exchanger type (e.g., tube and shell, fan-cooled)
- Heat exchanger duty rating
- Target or actual process stream discharge pressure
- Pressure drop across heat exchanger

5.2 Chokes

The following data is required for each choke to be modeled:

- Is the choke size adjustable or fixed? Note that for the purposes of the system optimization study, any choke size that can be changed by operator intervention is adjustable. In most chokes that are of "fixed choke" design, the bean size can be changed, and therefore they are adjustable by this definition.
- Choke size for a fixed choke. Maximum choke size for an adjustable choke.

5.3 Pressure Control Valves

The following data is required for each PCV to be modeled.

- The side of the valve on which pressure is controlled (e.g., upstream/downstream if the flow direction is fixed)

6.0 Production Data

Measured production data is required to calibrate the individual equipment models and the network model as a whole. In order to calibrate equipment models, data is required for a variety of operating conditions so that the model is tested and tuned over as much of the equipment's operating range as possible. It is also desirable that the production data provided allows the equipment to be tuned in isolation—i.e., the tuning of one item of equipment does not depend on the tuning of another.

In most practical cases, items of equipment can be isolated for calibration purposes if the inlet and outlet values of pressure and temperature, as well as the flow rate through the equipment, are known at a given point in time.

Although this may seem straightforward, upstream oil and gas production systems generally do not have a high density of pressure and temperature instruments, and flowmeters are even more thinly distributed. Therefore, it is highly likely that sufficient data to calibrate in isolation will only be available for a small proportion of the equipment items in the network and that those will be the more complex equipment types such as compressors, which tend to be instrumented more fully.

In order to calibrate the network model as a whole, production data is required from points that are spread out over the entire network for a single operating condition and, as near as possible, a single point in time. During calibration of the network as a whole, the interactions between the performances of individual items of equipment are addressed, and equipment items that have insufficient data to be tuned individually are tuned simultaneously. This process is iterative since adjusting one model will affect all others, and several attempts will be required before a reasonable match to the observed network performance will be achieved.

Note that data points used for individual equipment tuning can also be used to tune the network as a whole. For equipment tuning, the data values associated with an item of equipment only need to have been measured at the same time as each other. For network tuning, the data values need to have been measured at the same time as data values throughout the rest of the network.

6.1 Individual Items of Equipment

As mentioned previously, a better calibration of individual items of equipment can be achieved if sets of the data described in this section can be provided for several different conditions covering the equipment's operating range. Of the items of equipment considered, only wells, pipes, chokes, heat exchangers, and compressors can or need to be calibrated individually.

1. For each well, the production test history for at least the three months prior to the current date is required to verify the wellhead performance tables or well models. Each production test record will ideally consist of the following items:
    - Date
    - Gas, condensate, and water flow rates
    - Wellhead pressure and temperature
    - Choke size
    - Flowline inlet pressure and temperature
    - Flowline outlet pressure and temperature
    - Bottomhole pressure and temperature (only required if well models have been provided)
2. The following production data items are required for individual calibration of pipes, chokes, and heat exchangers:
    - Date
    - Inlet pressure

- Inlet temperature
- Outlet pressure
- Outlet temperature
- Throughput flow rate (gas, condensate, water)
- Choke size (for chokes only)

3. For simple compressor models, the following data is required:
   - Date
   - Inlet pressure
   - Inlet temperature
   - Outlet pressure
   - Outlet temperature
   - Throughput flow rate
   - Individual stage discharge pressures
   - Individual stage discharge temperatures
   - Inter-cooler discharge temperatures
   - Power used by compression train
4. For detailed compressor models, the following data is required in addition to that listed above for simple compressor models:
   - Date
   - Gas turbine heat rate curves (speed vs. horsepower generation)
   - Compressor performance curves (for different stages, if applicable)
   - Gear box specifications
   - Power turbine and compressor stage speeds
   - Fuel gas flow
   - Individual stage recycle flow rates
   - Date on which measurements were taken
   - Pressures, temperatures, and flow rates for individual items of equipment, as listed above
   - Wellhead pressures and temperatures. Individual well flow rates (gas, condensate, water), if known.
   - Pressures and flow rates at sales and delivery points
   - Heat exchanger discharge temperatures
   - Separator operating pressures, temperatures, inlet and outlet rates
   - PCV set-point pressures
   - FCV set-point rates

6.2 Entire Network

In order to calibrate the network as a whole, as many measurements as possible from all over the network are required for a single point in time. These will include:

- Date on which measurements were taken
- Pressures, temperatures, and flow rates for individual items of equipment, as listed above
- Wellhead pressures and temperatures. Individual well flow rates (gas, condensate, water), if known.
- Pressures and flow rates at sales and delivery points
- Heat exchanger discharge temperatures
- Separator operating pressures, temperatures, inlet and outlet rates
- PCV set-point pressures
- FCV set-point rates

7.0 Objectives and Constraints

In order to use the model to determine the optimum field configuration for both current and future operating conditions, it is necessary to enter objectives and constraints that will define the objective function that the optimizer is seeking to improve. In broad terms, the model constraints define the feasible range of operating rates and pressures at various points in the network, and the model objectives define rewards on rates or pressures that need to be maximized (such as gas sales) or penalties on rates or pressures that need to be minimized.

The following information is required:

7.1 Delivery and Sales Points

- Maximum or minimum permitted rates. These could be dictated either by the physical constraints of the system or by sales agreements.
- Maximum or minimum permitted pressures
- Price (per unit volume) of fluid leaving the network through the delivery/sales point
- If appropriate, change in price with delivery pressure
- Limits on the concentration of individual components (such as CO_2 or H_2S) and/or change in price with component concentration

7.2 Compressors

- For simple compressor models, the cost of compression per unit of power per day
- For detailed compressor models, any additional costs of compression over and above the loss of revenue from saleable gas being burnt as fuel
- For both types of models, limits on flow rates and pressures where these are not implicit in the model data described in Section as required

7.3 Other Equipment

- Minimum and maximum operating rates wherever there are physical constraints on these in the network. An example is separation vessel throughput.
- Minimum and maximum operating pressures wherever there are physical constraints on these in the network. An example is separation vessel pressure.

A3

Net Present Value

Summary

Net present value (NPV) is the difference between the present value of cash inflows and the present value of cash outflows over a period of time. NPV is used in capital budgeting to compare projects based on their expected rates of return, required investment, and anticipated revenue over time.

The concept of net present value uses the time value of money and is used to identify the project that gives the best return for an investment. A dollar today in hand has more value than the dollar that is expected to come after three years. This is because, with the money in hand now, the investor has the ability to earn possible returns through alternate investment channels.

Net present value makes use of all future cash flows, both negative and positive, discounted to the present value. The analysis is used to determine which project should be pursued when several options are available.

NPV Calculation Method

For the calculation, the amount of investment required today and all the returns that are expected in the future are accounted for. Stated otherwise, all the negative and positive cash flows are considered. Usually, the spending is taken as negative, and the income is treated as positive.

The net cash flow is calculated for Year 0, Year 1, Year 2, Year 3, and so on.

Using a discount rate, the net cash flow is converted to their present values. The discount rate represents a minimum rate of return acceptable to the investor (usually assumed to be the prevailing interest rate from a bank investment) or an assumed rate of inflation.

The present values are added up to get the NPV. The project option that gives the highest NPV will be the best investment option.

Formula for NPV

To calculate NPV, one needs to estimate future cash flows and discount to present values using the preferred discount rate.

If there is only one cash flow from a project that will be paid a certain number of years from now, the calculation for the NPV is as follows:

$$NPV = \frac{Cash\ flow}{(1+i)^t} - Initial\ Investment$$

Where

i = Required return or discount rate

t = Number of time periods

If the cash flows are at regular intervals – say, once a year – NPV can be calculated as follows:

$$NPV = \sum_{t=0}^{n} \frac{R_t}{(1+i)^t}$$

Where:

R_t = net cash inflow – outflows during a single period

i = discount rate or return that could be earned in alternative investments

t = Number of time periods.

In simple terms, NPV = today's value of the expected cash flows – today's value of invested cash.

Example 1

Option 1: Assume there is an investment of $50,000 now, and the total return promised is $75,000, to be paid as a lump sum three years from now. Considering a discount rate of 10%, the NPV is calculated as follows:

$$NPV = \frac{75000}{(1.1)^3} - 50000$$

NPV = 56348.61 – 50000 = $6,349

Option 2: Assume there is an investment of $50,000 now, and the total return promised is $75,000, paid as $15,000 per year for five years, starting from Year 1. Using a discount rate of 10%, the NPV is calculated as follows:

$$NPV = \frac{15000}{(1.1)^1} + \frac{15000}{(1.1)^2} + \frac{15000}{(1.1)^3} + \frac{15000}{(1.1)^4} + \frac{15000}{(1.1)^5} - 50000$$

NPV = -50000 + 13636 + 12397 + 11270 + 10245 + 9314 = $6,862

Note that a payment of $15,000 that would be received in Year 5 is worth only $9314 today, based on the discount rate applied. **Table 1** shows the NPV calculation for this case:

Table 1

Disc Rate (%)	10.00						
	Year 0	Year 1	Year 2	Year 3	Year 4	Year 5	**NPV**
Cash flow	-50000	15000	15000	15000	15000	15000	
Present value	-50000	13636	12397	11270	10245	9314	**6862**

Option 3: However, if another opportunity is available to invest the same $50,000 with the total return of $75,000, but paid as $25,000 per year for three years, payment starting from Year 1, the NPV is different. This is shown in **Table 2**.

Table 2

Disc Rate (%)	10.00				
	Year 0	Year 1	Year 2	Year 3	**NPV**
Cash flow	-50000	25000	25000	25000	
Present value	-50000	22727	20661	18783	**12171**

Both projects require the same initial investment, and the returns are also the same, but the third option generates the income faster than the second option, so it has a higher NPV. This would be the preferred investment option. In this case, the benefit comes from the time value of money.

If a non-realistic or higher discount rate is used, the project NPV may become negative. However, it is still possible to compare two projects based on the results.

Calculation of IRR

Example 1 shows an investment of $50,000 with a return of $75,000. Ignoring the time value of money, the amount gained is $25,000 from an investment of $50,000; thus, the rate of return (RoR) is 50%.

The rate of return using discounted cash flows is known as internal rate of return (IRR). The internal rate of return is a discount rate that makes the net present value of all cash flows from a particular project or investment equal to zero. In general, when comparing investment options with other similar characteristics, the investment with the highest IRR would be considered the best.

In the second option of Example 1 above, solving the NPV equation to derive the discount rate that yields zero value for NPV, the IRR can be obtained as 15.238%. This is shown in **Table 3**.

Table 3

IRR	15.238						
	Year 0	Year 1	Year 2	Year 3	Year 4	Year 5	**NPV**
Cash flow	-50000	15000	15000	15000	15000	15000	
Present value	-50000	13017	11295	9802	8506	7381	**0**

Using the same approach, the IRR for the third option of Example 1 is calculated as 23.375% (see **Table 4**).

Table 4

IRR	23.375				
	Year 0	Year 1	Year 2	Year 3	**NPV**
Cash flow	-50000	25000	25000	25000	
Present value	-50000	20263	16424	13312	**0**

This confirms that investment option 3 is better than option 2 in terms of returns.

Example 2

A simplified case of an economic evaluation for the selection of artificial lift method – gas lift vs. ESP, for a low-pressure onshore well – is presented in **Table 5**. The cost of installation is double for ESP in this case, but the *net* gain in production is 1.5 times more with ESP, compared to gas lift. The costs for well surveillance, maintenance, and fuel costs are already included in the incremental revenue. In other words, Incremental revenue = Oil gain – Maintenance costs. An annual oil production decline of 15% is assumed for the reservoir.

Table 5

Discount rate	8%							
	GL				**ESP**			
	Installation cost, $	**Incremental revenue, $**	**Net Cash flow, $**	**Present value, $**	**Installation cost, $**	**Incremental revenue, $**	**Net Cash flow, $**	**Present value, $**
Yr1	-100000	100000	0	0	-200000	150000	-50000	-50000
Yr2		85000	85000	78704		127500	127500	118056
Yr3		72250	72250	61943		108375	108375	92914
Yr4		61413	61413	48751	-200000	92119	-107881	-85640
Yr5		52201	52201	38369		78301	78301	57554
Yr6		44371	44371	30198		66556	66556	45297
NPV				**257965**				**178180**

In spite of the attractive returns for ESP, the NPV for the ESP option is lower than gas lift. This is mainly because of the cost of ESP replacement planned after three years, which is included in the cash flow.

Importance of Input Data

The cost of installation of gas lift assumes an existing gas lift infrastructure in the field. If this is not the case, the installation cost will increase, as a compressor would need to be purchased. This will decrease the NPV of gas lift. On the other hand, the rationale for using the same reservoir decline rate for both the lift options might be questionable. If gas lift increases the production rate by 100 BOPD and ESP by 150 BOPD, at 80% water cut, the increase of fluid rates would be 500 and 750 BLPD, respectively. Depending on reservoir parameters, the ESP lifted well may decline faster than the gas lifted well. In that case, the NPV for ESP will become smaller.

If there are more convincing technical or strategic reasons (example: accelerate production from a well anyway), the management may still opt for the ESP lift method. NPV gives the position based on the economic analysis of both options based on the best information available at the time of analysis.

Concerns

The NPV analysis works based on cash flow. The cash flow depends on several variables. Oil price is an important variable. The assumptions of production uplift from a project option and the associated reservoir decline are some key uncertainties. The analyst should have high confidence in the values used in the analysis and should follow a systematic approach to compare the results between two projects in any given situation.

List of Abbreviations

BLPD	Barrels of liquid per day (used synonymously with BFPD)
BOPD	Barrels of oil per day (volume at measured conditions of P and T)
BWPD	Barrels of water per day (volume at measured conditions of P and T)
CAPEX	Capital expenditure
CDS	Central degassing station
CGR	Condensate to gas ratio, STB/MMscf
DESC engr.	Design and evaluation services for clients engineer
DRA	Drag reducing agent
ESD	Emergency shut-down system
ESP	Electrical submersible pump
FAT	Field acceptance test
FBHP	Flowing bottom-hole pressure
FDS	Functional design specifications
FSG	Fluid source group
FWHP	Flowing wellhead pressure (same as FTP, flowing tubing-head pressure)
FWHT	Flowing wellhead temperature
GL	Gas lift
GLM	Gas lift mandrel
GLV	Gas lift valve
GOR	Gas-oil ratio, scf/STB Oil
GVF	Gas-void fraction (fraction of the channel volume that is occupied by the gas phase)
HP	High pressure
hp	Horsepower
HSE	Health, safety, and environmental
IAOM	Integrated asset optimization model
ICD	Inflow control device
ID	Internal diameter
IGOR	Injection gas-oil ratio, scf/STB Oil
IPM	Integrated production model
IRR	Internal Rate of Return
KPI	Key performance indicator

LP	Low pressure
MPP	Multi-phase booster pump
NF	Naturally flowing well
NPV	Net present value
OD	Outside diameter
OPEX	Operating expenditure
P	Pressure
P&ID	Piping and instruments diagram
PC	Personal computer
PFD	Process flow diagram
PI	Productivity index of a well, J (e.g., BLPD/psi drawdown)
PVT	Pressure-volume-temperature properties of fluids
Q	Flow rate
QA/QC	Quality assurance / Quality control
RDS	Remote degassing station
ROR	Rate of return (%)
RTO	Real-time production optimization
RTOC (D)	Real-time operations center (Drilling)
RTOC (P)	Real-time operations center (Production)
SCADA	Supervisory control and data acquisition system
SIBHP	Shut-in bottom-hole pressure
SLP	Sequential linear programming
SQP	Sequential quadratic programming
SSSV	Sub-surface safety valve
STB/Day	Stock tank barrels per day
T	Temperature
UAT	User acceptance test

Index

A

C

D

E

F

G

H

I

N

O

P

R

S

T

U

W